U.S. Food and Drug Administration

Methods for the Microbiological Analysis of Selected Nutrients

Gerald Angyal
Division of Science and Applied Technology
Office of Food Labeling

AOAC
INTERNATIONAL

The Scientific Association
Dedicated to Analytical Excellence®

1996

Published and Distributed by

AOAC INTERNATIONAL

481 North Frederick Avenue, Suite 500

Gaithersburg, MD 20877-2417

Telephone: +1-301-924-7077

Fax: +1-301-924-7089

Internet e-mail: aoac@aoac.org

Internet Web Site: http://www.aoac.org

ISBN 0-935584-61-7

CONTENTS

[The numbers for each chapter title (e.g., 300, 310, 320) are in-house standard operating procedure numbers and are included only to facilitate cross-referencing.]

Preface

This text is intended to provide analysts with methodology suitable for the microbiological analysis of specific nutrients. It contains standard operating procedures describing microbiological methods used by the U.S. Food and Drug Administration (FDA) in analyzing the water-soluble vitamins biotin, folic acid, niacin, pantothenic acid, riboflavin, thiamin, pyridoxine, vitamin B_{12}, and several other nutrients including choline and tryptophan. Many of the procedures are based on AOAC® Official Methods, which are identified for each nutrient. The procedure for the microbiological analysis of panthenol, an ingredient in many cosmetics, is also included.

A number of AOAC® Official Methods are available for analyzing some of the nutrients included in this manual. For example, AOAC microbiological and manual and semi-automated fluorometric methods are available for the analysis of riboflavin. However, AOAC currently does not have official microbiological methods for the analysis of thiamin, but instead uses thiochrome-based fluorometric procedures. The AOAC Task Force on Methods for Nutrient Labeling Analyses recommended that the *Lactobacillus viridescens* assay, a microbiological assay for thiamin, be collaboratively studied for all food matrices. Methods for niacin analysis include the microbiological procedure, which has been recommended for use in most food matrices, as well as colorimetric and high-performance liquid chromatographic (LC) procedures. Other microbiological AOAC® Official Methods include those for the analysis of pyridoxine, niacin, vitamin B_{12}, folate, and pantothenic acid in infant formula.

With the finalization of regulations requiring the addition of folic acid to certain cereal-grain products (U.S. Food and Drug Administration (March 5, 1996) *Fed. Regist.* **61**(44) pp 8781–8797), there is increased need for attention to rapid analysis methods of total folates in foods. Although microbiological assay methods for folate are currently AOAC® Official Methods, a number of LC methods for folate have been published (cf. Lambert, W. E. and DeLeenheer, A. (1992) *Food Analysis by HPLC*, Leo M. L. Nollet (Ed), Marcel Dekker, Inc., New York, NY, pp. 341–369).

To date, methods for folate analysis by liquid chromatography have not been collaboratively studied; the microbiological assays are the AOAC® Official Methods. Because LC methods are more rapid than microbiological assays, they offer the potential for developing folate assays that are more appropriate for quality control applications during food manufacturing or processing.

The interested reader is referred to *Methods of Analysis for Nutrition Labeling* [Sullivan, D. M. and Carpenter, D. E. (1993) AOAC INTERNATIONAL, Gaithersburg, MD] for additional information on the applicability of AOAC microbiological assays for specific nutrients. The text *Analyzing Food for Nutrition Labeling and Hazardous Contaminants* ((1995) Jeon, I. K. and Ikins, W. G. (Eds), Marcel Dekker, Inc., New York, NY, p. 496) provides up-to-date information describing currently available methods, including microbiological methods, for the analysis of many nutrients in foods.

We wish to acknowledge the assistance of Jeffrey S. Smith, Martin P. Bueno, and Ellen M. Anderson in developing this manual. Users of this manual may submit comments to Gerald Angyal, U.S. Food and Drug Administration, Office of Food Labeling, HFS-175, 200 C St., SW, Washington, DC 20204.

Jeanne I. Rader, Ph.D.
Director, Division of Science & Applied Technology
Office of Food Labeling

300 Microbiological Methods
General Information

(Not applicable in presence of extraneous turbidity or color that interferes with turbidimetric measurements.)

(Protect solutions from undue exposure to light throughout all stages, except where otherwise directed.)

A. Principle

Microbiological assays are based on a microorganism's requirements for nutritional substances such as minerals, amino acids, and vitamins that must be taken from the microorganism's environment. For microbial assays, the growth of the test organism in the assay solution is compared with that in a standard solution. The growth response is limited only by the nutrient under study. All other effects occur equally on the organism in the two solutions. In essence, the assay preparation behaves as if it were the standard preparation diluted by an inert substance.

Equivalent responses of the test organism at different concentration (potency) levels of the assay solution and at corresponding levels of the standard solution are a fundamental criteria of assay validity. This does not mean that the response curves are coincident, although they may be. Vitamin concentration in the assay solution is determined by direct or graphic interpolation on a concentration-response curve of a standard solution of the vitamin being investigated. The accuracy of the determination increases as the concentration of the assay solution approaches the concentration of the standard solution.

B. Variability of Test Organism Response

The assay methods provide concentration ranges for standard solutions and for dilutions of the assay solutions. The concentration is determined by finding the amount of standard necessary to generate the required bacterial response in a trial assay standard solution with each new lot of medium or culture of the microorganism. When the appropriate concentration has been established, the concentration-response curve is plotted with % transmittance on the y-axis and concentration on the x-axis revealing a practically parabolic curve. Additionally, the response at the 5.0 mL level of the standard will be equivalent to a dried cell weight ≥1.25 mg per tube.

The test organism may respond too vigorously, which is desired if the growth is not caused by a stimulant in the basal medium or by a nutrient carried over from the enriched culture medium. Occasionally, a basal medium contains a substance that will stimulate excessive growth of the test organism. This is the first thing to be suspected when a new lot of medium or a new medium component is used in an assay and the inoculated blank gives a transmittance reading equivalent to a dried cell weight >0.6 mg per tube. In such cases, it may be necessary to discard the medium, if it is a stock dry medium, or to determine which of the components contains the stimulatory substance if the medium is prepared by the analyst from basic chemicals.

Carry-over from enriched culture medium can be eliminated by washing the test organism two or three times with sterile saline solution or basal medium during preparation of the inoculum. If the test organism continues to grow vigorously after extra washings, its response can be reduced by decreasing the potency of the standard and assay solutions.

C. Glassware Cleanliness and Water

Glassware must be washed with nonstimulating detergents, such as sodium lauryl sulfate or 7X detergent, and rinsed free of detergent residues with distilled water and ethanol. Deionized water is not recommended for rinsing glassware or for making assay dilutions. Deionizers are frequently host to bacteria and the resins are slightly soluble. Culture tubes are heated to 250°C for at least 1 h to burn off any organic residues that might be present.

D. Instrumentation

(a) *Spectrophotometer.*—Bausch & Lomb Spectronic 20 or equivalent to read 20 × 150 mm (or 18 × 150 mm) test tubes.

(b) *pH meter.*—With long combination electrode.

(c) *Balance.*—To at least four decimal places.

(d) *Calculator.*—Calculations of assay results can also be accomplished using a computer (1) or a personal computer (2).

(e) *Optional automated assay tube reading.*—Gilson escargot fractionator Model SC-30 or Model 222 sample changer that has been modified with an air agitation system and is connected to a spectrophotometer with a flow cell and either a printer or a computer.

E. Apparatus

(a) *Test tubes.*—Disposable, 20 × 150 mm, and heated 1–2 h at 250°C.

(b) *Syringe.*—10 mL Luer-Lok with long 16 gauge needle.

(c) *Glassware.*—Graduated cylinders, volumetric flasks, Erlenmeyer flasks, beakers, volumetric pipets, watch glasses, and 20 × 150 screw top test tubes.

(d) *Test tube racks.*—Stainless steel, 4 × 12 to hold 20 × 150 mm tubes.

(e) *Large rack with cover to hold four or more test tube racks.*

(f) *Desiccator.*

(g) *Inoculating loops and straight wire.*

(h) *Bunsen burner.*

(i) *Pipetting machines.*—To deliver 1 and 5 mL aliquots. BBL Brewer (Cockeysville, MD) Models 93112 and 60453 or equivalent.

(j) *Autoclave.*—For sterilizing at 15 psi and 121–123°C.

(k) *Constant temperature water bath or incubator.*—With rotary shaker.

(l) *Centrifuge.*—To accommodate 20 × 150 mm test tubes.

(m) *Hot plate stirrer.*

(n) *Refrigerator and freezer.*

(o) *Vortex mixer.*

(p) *Hand pipet.*—Eppendorf repeater pipet or equivalent with a 50 mL capacity.

(q) *Adjustable digital pipet.*—Eppendorf (100–1000 μL) or equivalent.

F. Stock Basal Medium Solutions

[Store all solutions, except for solutions containing alcohol, in glass-stoppered bottles in refrigerator (2–8°C) with layer of toluene (3, section A).]

(a) *Vitamin-free acid-hydrolyzed casein.*—Commercial source. Humko Sheffield Chemical–Hy case amino or equivalent.

(b) *Adenine–guanine–uracil solution.*—Dissolve 1.0 g each of adenine sulfate, guanine hydrochloride, and uracil in 50 mL warm HCl (1 + 1), cool, and dilute with H_2O to 1 L.

(c) *Asparagine solution.*—Dissolve 10 g L-asparagine monohydrate in H_2O and dilute to 1 L.

(d) *Cystine solution.*—Suspend 2.0 g L-cystine in ca 750 mL H_2O, heat to 70–80°C, and add HCl (1 + 1) dropwise, stirring until solid dissolves. Cool and dilute with H_2O to 1 L.

(e) *Cystine–tryptophan solution.*—Suspend 8.0 g L-cystine and 2.0 g L-tryptophan (or 4.0 g D,L-tryptophan) in ca 1.5 L H_2O, heat to 70–80°C, and add HCl (1 + 1) dropwise, stirring until solids dissolve. Cool and dilute with H_2O to 2 L.

(f) *Manganese sulfate solution.*—Dissolve 5.0 g $MnSO_4 \cdot H_2O$ in H_2O and dilute to 500 mL.

(g) *Photolyzed peptone solution.*—Dissolve 100 g peptone (Difco Laboratories, Detroit, MI, No. 0118-02) in 625 mL H_2O, add solution of 50 g NaOH in 625 mL H_2O and mix in vessel (such as a crystallizing dish) so that solution is 1–2 cm deep. Place 100–500 watt bulb, fitted with reflector, ca 30–50 cm from solution. Expose solution to light from bulb, stirring occasionally and maintaining solution at ≥25°C, until riboflavin is destroyed (4–10 h is usually sufficient). Adjust solution to pH 6.0–6.5 with acetic acid, add 18 g anhydrous sodium acetate, stir until solid dissolves, dilute to 2 L with H_2O, and filter until solution is clear.

(**h**) *Polysorbate 80 solution.*—Dissolve 25 g polysorbate 80 (Tween 80) (polyoxyethylene sorbitan mono-oleate) in ethanol to final volume of 250 mL.

(**i**) *Salt solution A.*—Dissolve 50 g anhydrous KH_2PO_4 and 50 g anhydrous K_2HPO_4 in H_2O, dilute to 1 L, and add 10 drops HCl.

(**j**) *Salt solution B.*—Dissolve 20 g $MgSO_4 \cdot 7 H_2O$, 1.0 g NaCl, 1.0 g $FeSO_4 \cdot 7 H_2O$, and 1.0 g $MnSO_4 \cdot H_2O$ in H_2O, dilute to 1 L, and add 10 drops HCl.

(**k**) *Tryptophan solution.*—Suspend 2.0 g L-tryptophan (or 4.0 g D,L-tryptophan) in 700–800 mL H_2O, heat to 70–80°C, and add HCl (1 + 1) dropwise, stirring until solids dissolve. Cool and dilute with H_2O to 1 L.

(**l**) *Vitamin solution I.*—Dissolve 25 mg riboflavin, 25 mg thiamine hydrochloride, 0.25 mg biotin, and 50 mg niacin in 0.02 N acetic acid (1.2 mL acetic acid per liter) to final volume of 1 L.

(**m**) *Vitamin solution II.*—Dissolve 50 mg p-aminobenzoic acid, 25 mg calcium pantothenate, 100 mg pyridoxine hydrochloride, 100 mg pyridoxal hydrochloride, 20 mg pyridoxamine dihydrochloride, and 5 mg folic acid in 25% ethanol and dilute to 1 L.

(**n**) *Vitamin solution III.*—Dissolve 10 mg p-aminobenzoic acid, 40 mg pyridoxine hydrochloride, 4 mg thiamine hydrochloride, 8 mg calcium pantothenate, 8 mg niacin, and 0.2 mg biotin in ca 300 mL H_2O. Add 10 mg riboflavin dissolved in ca 200 mL 0.02 N acetic acid. Add solution containing 1.9 g anhydrous sodium acetate and 1.6 mL acetic acid into ca 40 mL H_2O. Dilute to 2 L with H_2O.

(**o**) *Vitamin solution IV.*—Dissolve 20 mg riboflavin, 10 mg thiamine hydrochloride, and 0.04 mg biotin in 0.02 N acetic acid to final volume of 1 L.

(**p**) *Vitamin solution V.*—Dissolve 10 mg p-aminobenzoic acid, 20 mg calcium pantothenate, and 40 mg pyridoxine hydrochloride in 25% ethanol to final volume of 1 L.

(**q**) *Vitamin solution VI.*—Dissolve 10 mg p-aminobenzoic acid, 50 mg niacin, and 40 mg pyridoxine hydrochloride in 25% ethanol to final volume of 1 L.

(**r**) *Xanthine solution.*—Suspend 1.0 g xanthine in 150–200 mL H_2O, heat to ca 70°C, add 30 mL NH_4OH (2 + 3), and stir until solid dissolves. Cool and dilute to 1 L with H_2O.

(**s**) *Yeast supplement solution.*—Dissolve 20 g H_2O soluble yeast extract (Difco, No. 0127-02) in 100 mL H_2O, add turbid solution of 30 g Pb subacetate in 100 mL H_2O, and mix. Filter and adjust filtrate to pH 10 with NH_4OH (1 + 2). Filter and adjust filtrate to pH 6.5 with acetic acid. Precipitate excess Pb with H_2S, filter, and dilute filtrate to 200 mL with H_2O.

(**t**) *Vitamin solution VII.*—Dissolve 10 mg p-aminobenzoic acid, 10 mg calcium pantothenate, 10 mg niacin, and 20 mg pyridoxine hydrochloride in 25% ethanol to final volume of 1 L.

(**u**) *PABA-Vitamin B_6 solution.*—Dissolve 50 mg p-aminobenzoic acid and 120 mg pyridoxine hydrochloride in 200 mL H_2O. Add 0.95 g sodium acetate and 0.8 mL acetic acid to ca 40 mL H_2O. Combine two solutions and dilute to 500 mL with H_2O.

(**v**) *Norit A treated casitone solution.*—Dissolve 100 g Difco casitone (No. 0259-01) in H_2O by stirring and dilute to 1 L. Steam 10 min and cool. Adjust pH to 1.5 with concentrated HCl. Shake 20 min with 70 g Norit A (carbon decolorizing alkaline, Fisher Scientific Co. [No. C-176]) and filter. Adjust filtrate to pH 1.5 with HCl, repeat Norit A treatment, and filter. Adjust filtrate to pH 6.0, steam 10 min, cool, and filter. Store under toluene in refrigerator.

(**w**) *Liver extract solution.*— Dissolve 5 g Difco liver (No. 0133-01) in 100 mL H_2O. Heat for 1 h at 50°C with frequent agitation. Boil mixture for a few minutes to coagulate a portion of the proteins, cool, dilute to 100 mL, and filter. Dispense 5 mL in 20 × 150 mm screw top test tubes, cap, and autoclave 15 min at 121–123°C. Cool and store in refrigerator. Contains ca 8 mg liver extract per mililiter.

(**x**) *Amino acid and vitamin mixture.*—Weigh 2 g b-alanine, 200 mg niacin, 200 mg p-aminobenzoic acid, 2 g thiamine, 2 g riboflavin, 2 g pyridoxine, 20 mg folic acid, and 16 mg biotin. Add glucose to a final weight of 100 g and mix thoroughly.

G. Culture and Suspension Media

(3, section B)

(**a**) *Liquid culture media.*—Difco *Lactobacilli* broth has been found satisfactory. Add 10 mL portions of solution to screw top 20 × 150 mm test tubes, cover to prevent contamination, sterilize 15 min in autoclave at 121–123°C, and store in refrigerator immediately after heating.

(**b**) *Agar culture media.*—Difco *Lactobacilli* agar AOAC has been found satisfactory with the following additions: 3 g Difco agar per liter and 1 μg B_{12} solution per liter. After agar dissolves, add 10 mL hot solution to 20 × 150 mm test tubes, plug with cotton, and cap. Cover tubes to prevent contamination, sterilize 15 min at 121–123°C, and store in refrigerator upon completion of cycle.

(**c**) *Sterile saline.*—Dissolve 9 g NaCl in 1 L distilled H_2O. Add 10 mL portions to 20 × 150 mm test tubes capped with plastic top, sterilize 15 min at 121–123°C, and store in refrigerator upon completion of cycle.

H. Stock Cultures of Test Organisms

For appropriate test organism, prepare one fresh test tube stab or slant culture weekly [use **300.00B(b)**] or appropriate stab or slant as indicated in method. Incubate ca 19–22 h at 37°C unless stated otherwise. If organism is slow-growing, subculturing can be done more frequently until growth returns to normal (3, section C).

(**a**) *Lactobacillus leichmannii (ATCC 7830).*—For use in assay of vitamin B_{12} (cobalamin).

(**b**) *Streptococcus faecalis (ATCC 8043).*—For use in assay of free folic acid.

(**c**) *Lactobacillus plantarum (ATCC 8014).*—For use in assay of niacin, biotin, tryptophan, and free and bound pantothenic acid.

(**d**) *Lactobacillus casei rhamnosus (ATCC 7469).*—For use in assay of riboflavin and free and bound folic acid. Transfer culture through specific liquid culture media then back to agar slabs for increased specificity.

    (**e**) *Lactobacillus viridescens (ATCC 12706).*—For use in assay of thiamine.

    (**f**) *Saccharomyces uvarum (ATCC 9080).*—For use in assay of vitamin B_6.

    (**g**) *Neurospora crassa (ATCC 9278).*—For use in assay of choline.

    (**h**) *Acetobacter suboxydans (ATCC 621H).*—For use in assay of Panthenol.

I. Calibration of Photometer

Calibration is needed for each individual vitamin assay on the specific photometer to be used (3, section F).

    (**a**) Aseptically, add 1 mL inoculum for appropriate vitamin assay to 300 mL sterile basal medium containing 1.0 mL standard stock solution. Incubate mixture for same period and at same temperature to be employed in determination.

    (**b**) After incubating, centrifuge and wash cells 3× with ca 50 mL portions 0.9% NaCl solution; then resuspend cells in the NaCl solution to a final volume of 25 mL. Evaporate 10 mL aliquot of cell suspension on steam bath and dry to constant weight at 110°C in vacuum oven. Correcting for weight of NaCl, calculate dry weight of cells in mg/mL of suspension.

    (**c**) Dilute second measured aliquot of cell suspension with 0.9% NaCl solution so that each mL is equivalent to 0.5 mg dry cells. To test tubes add, in triplicate, 0.0, 0.5, 1.0, 1.5, 2.0, 2.5, 3.0, 4.0, and 5.0 mL, respectively, of diluted cell suspension.

    (**d**) To each tube, add 0.9% NaCl solution to a final volume of 5.0 mL. Add 5.0 mL appropriate basal medium, mix (1 drop of suitable antifoam agent may be added; 1–2% solution of Dow Corning antifoam AF emulsion or antifoam B has been found satisfactory), and read. With blanks set at 100% transmittance (%T), measure %T of each tube under same conditions to be used in respective assay. Prepare curve by plotting %T reading for each level of diluted cell suspension against cell content (milligrams dry cell weight) of respective tubes.

    (**e**) Repeat appropriate calibration step at least twice more for photometer to be used in respective assay.

    (**f**) Draw composite curve, best representing three or more individual curves, relating %T to milligrams dried cell weight for photometer under conditions of respective assay. Once appropriate curve for particular instrument is established, all subsequent relationships between %T and cell weight are determined directly from this curve. Respective assay limits are expressed as milligrams dried cell weight per tube.

J. Assay Procedure

    (3, sections D and G)

    (**a**) Prepare 20 × 150 mm disposable test tubes containing appropriate standard solution as follows: Add, in triplicate, two sets at 0.0 mL (one set for uninoculated blanks and one set for inoculated blanks), and one set at 1.0, 2.0, 3.0, 4.0, and 5.0 mL standard, respectively, using Brewer automatic pipetting machine calibrated to deliver 1 mL.

Prepare three check tubes of primary standard solution at 0.0 mL (for inoculated blanks) and 5 mL to determine if maximum growth has been obtained for the assay. To each tube of standard solution, add H_2O for a total volume of 5 mL using Brewer automatic pipetting machine. Add 5.0 mL of appropriate double-strength basal medium to each tube using Brewer automatic pipetting machine calibrated to deliver 5 mL. Repeat procedure for secondary standards except no blanks or check tubes are necessary.

(b) Prepare 20×150 mm disposable test tubes containing appropriate sample assay solution as follows using Brewer automatic pipetting machine: Add, in duplicate, 1.0, 2.0, 3.0, and 4.0 mL sample assay solution. To each tube of sample assay solution add H_2O for a total volume of 5 mL. Add 5.0 mL of appropriate double-strength basal medium to each tube.

(c) Cover tubes with stainless steel top to prevent bacterial contamination, and autoclave 5 min at 121–123°C. Cool tubes as rapidly as possible to minimize reaction between amino acids and sugars in the basal medium. The products of this reaction will darken the color of the medium, and complex amino acids will not be available to the test organism. Take precautions to keep sterilizing and cooling conditions uniform throughout assay. Packing the tubes too closely in the autoclave or overloading it may cause variation in heating or cooling.

(d) Aseptically inoculate each tube, except one set of uninoculated blank tubes containing 0.0 mL standard solution, with 1 drop appropriate inoculum delivered from sterile 10 mL Luer-Lok syringe fitted with long 16 gauge needle. Drop must fall onto the surface of the tube contents. Needle must be held at same angle for inoculation of all tubes so that drop size will be constant. The tubes must not be warmer than 40°C when inoculum is added or the microorganism may be inactivated, negating the possibility of growth.

(e) Incubate for time period designated appropriate for method at selected temperature held constant (± 0.5°C) using constant temperature circulating H_2O bath or incubator. To determine when maximum growth has been obtained, remove one of the check tubes at 0.0 mL level and one check tube at 5.0 mL primary standard level and add one drop of antifoam to each. Mix and read on spectrophotometer at 550 nm using the inoculated blank (0.0 mL) for 100%T and observe %T of the 5.0 mL check standard tube. After 2 h, repeat with two more tubes. A change of >1–3%T indicates the assay has completed growing. If there is >3% change, then last two check tubes should be read at the next 2 h interval not to exceed 24 h total incubation time.

(f) After the assay has reached full growth, add 1 drop of suitable antifoam agent solution (1–2% solution of Dow antifoam AF emulsion or antifoam B is satisfactory) to each tube and agitate. Place tube in photometer (for calibration of assay see section I) set at 550 nm and read %T when steady state is reached. Reading time or steady state is dependent on the characteristics of the test organism. If steady state is not reached within a reasonable time, read each tube at the same time interval so readings are reproducible. With transmittance set at 100%T for uninoculated blank level, read %T of inoculated blank level. Uninoculated blank level should be clear, if not, either the tubes or surroundings were dirty or the sterilization temperature was not attained. If inoculated blank reading (%T) corresponds to a dried cell weight >0.6 mg per tube (for calibration of photometer *see* section **I**), disregard results of assay. High inoculated blanks may be due to contamination of basal medium or culture. Mix inoculated blank tubes and reset transmittance to 100%T. Read all standard tubes. If the 5.0 mL level reading (%T) corresponds to a dried cell weight of <1.25 mg per tube (for calibration of photometer *see* section **I**), disregard results of assay. Read all sample tubes.

K. Calculations

(3, section H)

(**a**) Prepare standard concentration-response curve by plotting the average %T readings for each level of standard solution used against amount of reference standard contained in respective tubes. Determine the amount of vitamin per milliliter for each tube in each level of sample assay solution by interpolation from standard curve. Discard any observed %T value of sample assay solution equivalent to <0.5 mL or >4.5 mL, respectively, of standard solution. Average these values obtained from assay tubes and find a range ±10% of the average. Accept tubes within range, and if number of acceptable values remaining is ≥⅔ of the original number of tubes used in the four levels of sample assay solution, calculate potency of sample from the average of acceptable tubes.

(**b**) The program for computerized calculation of results in applied program language (APL) is found in (1).

(**c**) The program for computerized calculation of results in Basic for personal computers is found in (2).

References

(1) Broulund, G. V., Haskins, E. W., and Hudson, G. A. (1973) *J. Assoc. Off. Anal. Chem.* **56**, 754–757.

(2) Anderson, E. M. (1989) *J. Assoc. Off. Anal. Chem.* **72**, 950–952.

(3) *Official Methods of Analysis* (1995) 16th Ed., AOAC INTERNATIONAL, Gaithersburg, MD, method **960.46**.

310 Biotin

A. Basal Medium

(**a**) *Basal medium.*—Biotin-free double-strength. Prepared using commercially available Difco biotin assay medium (No. 0419-15) as directed.

B. Standard Solutions

(**a**) *Stock solution I (100 μg/mL).*—Weigh 50 mg anhydrous D-biotin (free acid), previously dried over silica gel under vacuum, and dilute to 500 mL with 25% ethanol.

(**b**) *Stock solution II (1 μg/mL).*—Dilute 5 mL stock standard solution I to 500 mL with 25% ethanol.

(**c**) *Intermediate solution (10 ng/mL)*—Dilute 5 mL stock solution II to 500 mL with 25% ethanol. This solution must be prepared fresh monthly.

(**d**) *Working solution.*—Prepare day of assay. Dilute 5 mL intermediate solution to 500 mL with H_2O (0.1 ng/mL). This is the primary standard with an equivalent assay rate to the samples. Two secondary standards are also run, one higher (usually 0.14 ng/mL) and one lower (usually 0.08 ng/mL) than the primary standard, in case the sample is either too high or low for the primary standard to be used.

C. Inoculum

(**a**) *Test organism.*—*Lactobacillus plantarum* (ATCC 8014) maintained through weekly transfers on agar maintenance medium incubated for 16–21 h at 37°C. Store in refrigerator.

(**b**) *Agar maintenance medium.*—Use **300G(b)**.

(**c**) *Liquid culture medium.*—Dilute a measured volume of basal medium with an equal volume of H_2O containing 0.1 ng biotin per mililiter. Add 10 mL portions of diluted medium to 20×150 mm screw top test tubes, autoclave 15 min at 121–123°C, and cool tubes as quickly possible. Store in refrigerator.

(**d**) *Working inoculum.*—Transfer the day before assay is being run. Transfer cell from stock culture of *L. plantarum* to two sterile tubes containing the liquid culture medium, keeping all transfers as sterile as possible. Incubate 16–21 h in a constant temperature bath at 37°C. Under aseptic conditions, centrifuge the culture and decant the supernatant. Wash the cells with 10 mL sterile saline **300G(c)**. Repeat procedure two more times. After the third washing with sterile saline, resuspend the cells. Cell suspension is the inoculum.

D. Assay Solution

(a) Accurately weigh a sample containing 200 ng biotin and transfer to a 250 mL beaker.

(b) Add ca 100 mL 1 N H_2SO_4 and disperse sample evenly in the liquid.

(c) Autoclave mixture for 30 min at 121–123°C and cool.

(d) Adjust mixture to pH 4.5 with NaOH.

(e) Dilute mixture to first serial dilution with H_2O and filter.

(f) Take known aliquot of the clear filtrate, check for complete precipitation with dilute HCl, and proceed as follows: *(1)* If no further precipitation occurs, adjust the clear filtrate to pH 6.8 with NaOH. Dilute with H_2O to a measured volume containing 0.1 ng biotin per mililiter. *(2)* If further precipitation occurs, adjust the mixture again to a point of maximum precipitation, dilute to a known volume with H_2O, filter, and take a known aliquot of the clear filtrate. Adjust pH to 6.8 with NaOH and dilute with H_2O to a measured volume containing 0.1 ng biotin per mililiter.

E. Assay

Using standard solution, **310B(d)**; assay solution, **310D**; basal media, **310A**; and inoculum, **310C(d)**; proceed as in **300J** and **K**.

References

(1) *Analytical Microbiology* (1963), F. Kavanaugh, (Ed.), Academic Press, pp. 421–428.

(2) Wright, L. D. and Skeggs, H. R. (1944) *Proc. Soc. Exp. Biol. Med.* **56**, 95–98.

(3) *Official Methods of Analysis* (1995) 16th Ed., AOAC INTERNATIONAL, Gaithersburg, MD, method **960.46D, G,** and **H**.

320 Choline

A. Basal Medium

The basal medium can be prepared by using commercially available Difco choline assay medium (No. 0460-15).

B. Standard Solutions

(**a**) *Stock solution (500 μg/mL).*—Accurately weigh 576 mg choline chloride that has been dried to constant weight. Dissolve in H_2O, dilute to 1 L, and store in refrigerator (2–8°C).

(**b**) *Working solutions.*—Prepare day of assay. Dilute 5 mL stock solution to 500 mL with H_2O (5.0 μg/mL). This is the primary standard with an equivalent assay rate to the sample.

Also run two secondary standards, one higher (usually 7.0 μg/mL) and one lower (usually 3.0 μg/mL) than the primary standard, in case the sample is either too high or too low for the primary standard to be used.

C. Inoculum

(**a**) *Test organism.*–*Neurospora crassa* (ATCC 9277) maintained through weekly transfers on agar maintenance medium incubated for 16–21 h at 30°C. Store in refrigerator.

(**b**) *Agar maintenance medium.*—Dissolve 0.4 g Difco yeast extract (No. 0127-02), 0.4 g Difco malt extract (No. 0186-01), and 1.5 g Difco agar (No. 0140-01) in 100 mL H_2O. Add 100 μg of choline to solution and heat to boil. Pipet hot agar in 7 mL portions into 20 × 150 mm test tubes, plug with cotton, cap, and autoclave 15 min at 121–123°C. After autoclaving, tilt hot agar tubes to form slants and cool in this position. Store in refrigerator. Difco *neurospora* agar (No. 0321-15) has been found satisfactory as a commercially available medium.

(**c**) *Working inoculum.*—Transfer cells from *N. crassa* stock culture to a sterile tube containing 10 mL of saline [**300G(c)**]. Suspension obtained is inoculum.

D. Assay Solution

(**a**) Accurately weigh sample containing ≥1.0 mg of choline and transfer to a 250 mL beaker.

(**b**) Add ca 100 mL of 3% H_2SO_4, autoclave 2 h at 121–123°C, and cool.

(**c**) Adjust mixture to pH 4.5 with saturated barium hydroxide, dilute to measured volume, and filter using Celite if necessary.

(**d**) Adjust an aliquot of the clear filtrate to pH 5.5 with anhydrous $Ba(OH)_2$ and dilute with H_2O to a measured volume containing ca 5 µg choline per mililiter assay rate concentration.

(*Note*: D,L-methionine interferes with the choline assay when present in excess of 4.0 µg/mL. To remove methionine, pass solution through a 110 × 0.6 mm glass column containing an ion-exchange resin. AG 50W-X8 [BioRad] has been found satisfactory. Use 10 mL of 5% NaCl solution to elute choline.)

E. Assay

(**a**) *Pipetting and inoculating.*—Samples and standards are distributed among six 125 mL Erlenmeyer flasks in duplicate (0.0, 1.0, 2.0, 3.0, 4.0, 5.0 mL) with the concentration of choline ranging from 0–25 µg. Add H_2O to flasks for a total volume of 10 mL. Add 10 mL of basal medium to each flask. Autoclave 5 min at 121–123°C and cool. Add 1 drop of inoculum per flask and incubate 72 h at 25–30°C.

(**b**) *Reading the assay.*—At end of the incubation period, steam flasks for 5 min at 100°C. Remove all mycelium from each flask using an inoculating loop, press dry between paper towels or equivalent, and roll into a small pellet. Dry pellets 2 h at 100°C in a vacuum oven. A glazed porcelain spot plate is convenient for handling mycelium during drying and weighing. Weigh dried pellets to the nearest 0.5 mg.

F. Calculations

A standard curve is constructed from the weights obtained and the unknown determined by interpolation (weight vs. concentration).

Reference

(1) Horowitz, N. H. and Beadle, G. W. (1943) *J. Biol. Chem.* **150**, 325–333.

330 Folic Acid
AOAC

(Applicable to products containing free folic acid.)

A. Basal Medium

(a) Folic acid-free double-strength basal medium.—(Stock basal medium solutions, *see* **300F**).—

mL of basal medium to prepare	250	500	1000
(Add in order listed) (mL)			
Tryptophan solution [**300F(k)**]	25	50	100
Adenine–guanine–uracil solution [**300F(b)**]	2.5	5	10
Xanthine solution [**300F(r)**]	5	10	20
Asparagine solution [**300F(c)**]	15	30	60
Vitamin solution III [**300F(n)**]	50	100	200
Salt solution B [**300F(j)**]	5	10	20

Add ca 100 mL H_2O and the following solids (g):			
Vitamin-free casein, hydrolyzed	2.5	5	10
Dextrose, anhydrous	10	20	40
Sodium citrate·$2H_2O$	13	26	52
Potassium phosphate dibasic	1.6	3.2	6.4
L-cysteine·HCl·H_2O	0.19	0.38	0.76
Glutathione	.00013	.00026	.00052

Mix, adjust to pH 6.8 with NaOH, and add the following (mL):			
Polysorbate 80 solution [**300F(h)**]	0.25	0.5	1
Manganese sulfate solution [**300F(f)**]	5	10	20

Dilute to volume with H_2O

(b) The basal medium can also be prepared by using commercially available Difco folic medium AOAC (No. 0967-15) as directed.

13

B. Standard Solutions

(Use dark bottles.)

(a) *Stock solution (100 µg/mL).*—Accurately weigh 50 mg U.S. Pharmacopoeia (USP) folic acid that has been dried to constant weight and suspend in ca 30 mL 0.01 N NaOH. Add 300 mL H_2O and adjust pH to 7–8 with HCl. Dilute to 500 mL with H_2O and store in refrigerator under toluene.

(b) *Intermediate solution I (1 µg/mL).*—Dilute 5 mL stock solution with H_2O and adjust to pH 7–8 with HCl. Dilute to 500 mL with additional H_2O and store in refrigerator under toluene.

(c) *Intermediate solution II (100 ng/mL).*—Prepare on day of assay. Dilute 50 mL of intermediate solution I to 500 mL with H_2O.

(d) *Working solutions.*—Prepare day of assay. Dilute 5 mL of intermediate solution II to 500 mL with H_2O (1.0 ng/mL). This is the primary standard with an equivalent assay rate to the sample. Also run two secondary standards, one higher (usually 1.4 ng/mL) and one lower (usually 0.8 ng/mL) than the primary standard in case the sample is either too high or too low for the primary standard to be used.

C. Inoculum

(a) *Test organism.*—*Streptococcus faecalis* (ATCC 8043) maintained through weekly transfers on agar maintenance medium incubated for 16–21 h at 37°C. Store in refrigerator.

(b) *Agar maintenance medium.*—Use **300G(b)**.

(c) *Liquid culture medium.*—Dilute a measured volume of basal medium with an equal volume of H_2O containing 2.0 ng folic acid per mililiter. Add 10 mL portions of diluted medium to 20 × 150 mm screw cap test tubes, autoclave 15 min at 121–123°C, and cool tubes rapidly. Store in refrigerator.

(d) *Working inoculum.*—Transfer the day before assay is run. Transfer cells from stock culture of *S. faecalis* to two sterile tubes containing the liquid culture medium. Keep all transfers as sterile as possible. Incubate in a constant temperature bath for 16–21 h at 37°C. Under aseptic conditions, centrifuge the culture and decant the supernatant. Wash the cells with 10 mL sterile saline **300G(c)**. Repeat procedure two more times. After third washing with sterile saline, resuspend the cells. Cell suspension obtained is the inoculum.

D. Assay Solution

(a) Accurately weigh a sample containing ≤100 mg folic acid and transfer to a 250 mL beaker.

(b) Add volume of H_2O equal in mL to 10 times dry weight of sample. Resulting solution must contain ≤1.0 mg/mL folic acid (100 mL is minimum volume).

(c) Add equivalent of 2 mL NH$_4$OH (2 + 3) per 100 mL liquid. Autoclave mixture 15 min at 121–123°C and cool.

(d) Adjust mixture to pH 4.5 with NaOH, dilute to first serial dilution with H$_2$O, and filter.

(e) Take a known aliquot of the clear filtrate, check for complete precipitation with dilute HCl, and proceed as follows: *(1)* If no further precipitation occurs, adjust the clear filtrate to pH 6.8 with NaOH and dilute with H$_2$O to a measured volume containing 1.0 ng folic per mililiter. *(2)* If further precipitation occurs, adjust the mixture again to a point of maximum precipitation, dilute to a known volume with H$_2$O, and filter. Take known aliquot of clear filtrate, adjust to pH 6.8 with NaOH, and dilute with H$_2$O to a measured volume containing 1.0 ng folic per mililiter.

E. Assay

Using standard solution [**330B(d)**], assay solution (**330D**), basal media (**330A**), and inoculum [**330C(d)**], proceed as in **300J** and **K**.

References

(1) *Official Methods of Analysis* (1995) 16th Ed., AOAC INTERNATIONAL, Gaithersburg, MD, method **944.12H**.

331　Folic Acid

Free Form

(Applicable to products containing free folic acid.)

A. Basal Medium

(a) *Folic acid-free double-strength basal medium.*—(Stock basal medium solutions, *see* **300F**).

mL of basal medium to prepare	250	500	1000
(Add in order listed) (mL)			
Adenine–guanine–uracil solution [**300F(b)**]	2.5	5	10
Xanthine solution [**300F(r)**]	5	10	20
Asparagine solution [**300F(c)**]	15	30	60
Vitamin solution III [**300F(n)**]	50	100	200
Salt solution B [**300F(j)**]	5	10	20
PABA-Vitamin B_6 solution [**300F(u)**]	2.5	5	10

Add ca 100 mL H_2O and the following solids (g):			
Vitamin-free casein, hydrolyzed	2.5	5	10
Dextrose, anhydrous	10	20	40
Potassium phosphate dibasic	0.25	0.5	1
Potassium phosphate monobasic	0.25	0.5	1
Sodium acetate·$3H_2O$	16.6	33.2	66.4
Glutathione	.00125	.0025	.005

Dissolve the following solids (g) in dilute HCl and add to above solution:			
L-cysteine·HCl	0.125	0.25	0.5
D,L-tryptophan	0.05	0.1	0.2

Mix, adjust to pH 6.8 with NaOH and add the following (mL):			
Polysorbate 80 solution [**300F(h)**]	0.25	0.5	1

Dilute to volume with H_2O.

(**b**) The basal medium can also be prepared by using commercially available Difco folic acid casei medium (No. 0822-15) as directed.

B. Standard Solutions

(Use dark bottles.)

(**a**) *Stock solution (100 μg/mL).*—Accurately weigh 50 mg USP folic acid dried to constant weight and suspend in ca 30 mL 0.01 N NaOH. Add 300 mL H_2O and adjust to pH 7–8 with HCl. Dilute to 500 mL with H_2O. Store in refrigerator under toluene.

(**b**) *Intermediate solution (1 μg/mL).*—Dilute 5 mL stock solution with H_2O and adjust to pH 7–8 with HCl. Dilute to 500 mL with additional H_2O and store in refrigerator under toluene.

(**c**) *Working solution.*—Prepare day of assay. Dilute 5 mL of intermediate solution to 50 mL with H_2O (10 ng/mL). Dilute 10 mL (10 ng/mL) to 500 mL with H_2O (0.2 ng/mL). This primary standard with an assay rate equivalent to the samples. In addition, run two secondary standards, one higher (usually 0.3 ng/mL) and one lower (usually 0.12 ng/mL) than the primary standard in case the sample is either too high or too low for the primary standard to be used.

C. Inoculum

(**a**) *Test organism.*—*Lactobacillus casei* (ATCC 7469) maintained through weekly transfers on agar maintenance medium incubated for 16–21 h at 37°C. Store in refrigerator.

(**b**) *Agar maintenance medium.*—Use **300G(b)**.

(**c**) *Liquid culture medium.*—Prepare solubilized liver solution (1 mg/mL) by suspending 0.1 g Difco liver (No. 0133-01) in 100 mL H_2O. Hold mixture for 1 h at 50°C, filter, and store under toluene in refrigerator. Dilute a measured volume of basal medium with an equal volume of H_2O containing 0.8 ng folic acid and 20 μg solubilized liver per mililiter. Add 10 mL portions of diluted medium to 20 × 150 mm screw cap test tubes, autoclave 15 min at 121–123°C, and cool tubes rapidly. Store in refrigerator. Difco bacto inoculum broth (No. 0320-02) has been found satisfactory for liquid culture medium.

(**d**) *Working inoculum.*—Transfer the day before assay is to be run. Transfer cells from stock culture of *L. casei* to two sterile tubes containing the liquid culture medium. Keep all transfers as sterile as possible. Incubate 16–21 h in constant temperature bath at 37°C. Under aseptic conditions, centrifuge the culture and decant the supernatant. Wash the cells with 10 mL sterile saline [**300G(c)**]. Repeat procedure two more times. Resuspend cells after the third washing with sterile saline. Add 5 drops of suspended cells to 10 mL of sterile saline. Cell suspension obtained is the inoculum.

D. Assay Solution

(**a**) Prepare 1.42% Na_2HPO_4 solution with H_2O. Immediately prior to use, add 1 g ascorbic acid per 100 mL 1.42% Na_2HPO_4 solution and adjust pH to 7.8 with 4 N NaOH. Thirty mililiters of this extracting buffer is needed for each sample or standard.

(**b**) Accurately measure an amount of sample equal to 0.25–1.0 g dry solids and containing 1–5 µg total folates.

(**c**) Transfer sample to 125 mL Erlenmeyer flask containing 10 mL buffer and mix thoroughly.

(**d**) Add an additional 10 mL of buffer. Add 0.1 mL of octanol as antifoaming agent.

(**e**) Cover flask with beaker, autoclave 15 min at 121–123°C, cool, and add an additional 10 mL of buffer.

(**f**) Adjust samples to pH 4.5 with HCl, dilute to a measured volume with H_2O, and filter.

(**g**) Take known aliquot of the clear filtrate and dilute to final volume. The final solution will have a folic acid concentration of 0.2 ng/mL (assay rate) and will contain 2 mL pH 6.8 buffer per 100 mL. If final dilution step requires >10 mL aliquot, match the amount taken with pH 6.8 buffer. Preparation of the pH 6.8 buffer is described in **332E(c)***(2)*.

E. Assay

Using standard solution [**331B(c)**], assay solution (**331D**), basal media (**331A**), and inoculum [**331C(d)**], proceed as in **300J** and **K**.

References

(1) *Official Methods of Analysis* (1995) 16th Ed., AOAC INTERNATIONAL, Gaithersburg, MD, methods **944.12** (modified), **960.46D, G,** and **H**.

(2) Ford, J. E., Salter, D. N., and Scott, K. J. (1969) *J. Dairy Res.* **36**, 435–446.

332 Total Folates

Single Enzyme (AOAC Infant Formula)

This procedure is a modification of AOAC® Official Method **944.12H** (1). The extraction procedure is from Ford (2) and the assay procedure is from Flynn (3) and USDA Handbook No. 29 (4).

A. Principle

Folic acid (pteroylglutamic acid) occurs naturally in foods bound to glutamic acid residues of varying chain lengths. Most of these bound forms cannot be utilized by the microorganism and must be liberated prior to analysis. Folates are hydrolyzed to folic acid by conjugase enzyme from natural sources such as chicken pancreas. The liberated folic acid is then assayed using the conventional microbiological method.

B. Basal Medium

(**a**) *Folic acid-free double-strength basal medium.*—(Stock basal medium solutions, *see* **300F**).

mL of basal medium to prepare (Add in order listed) (mL)	250	500	1000
Adenine–guanine–uracil solution [**300F(b)**]	2.5	5	10
Xanthine solution [**300F(r)**]	5	10	20
Asparagine solution [**300F(c)**]	15	30	60
Vitamin solution III [**300F(n)**]	50	100	200
Salt solution B [**300F(j)**]	5	10	20
PABA-vitamin B_6 solution [**300F(u)**]	2.5	5	10
Add ca 100 mL H_2O and the following solids (g):			
Vitamin-free casein, hydrolyzed	2.5	5	10
Dextrose, anhydrous	10	20	40
Potassium phosphate dibasic	0.25	0.5	1
Potassium phosphate monobasic	0.25	0.5	1
Sodium acetate·$3H_2O$	16.6	33.2	66.4
Glutathione	.00125	.0025	.005

mL of basal medium to prepare	250	500	1000

(Add in order listed)

Dissolve the following solids (g) in dilute HCl and add to above solution:

	250	500	1000
L-cysteine·HCl	0.125	0.25	0.5
D,L-tryptophan	0.05	0.1	0.2

Mix, adjust to pH 6.8 with NaOH and add the following (mL):

Polysorbate 80 solution [**300F(h)**]	0.25	0.5	1

Dilute to volume with H_2O.

(b) The basal medium can also be prepared by using commercially available Difco folic acid casei medium (No. 0822-15, Difco Laboratories, Detroit, MI) as directed.

C. Standard Solutions

(Use dark bottles.)

(a) *Stock solution (100 µg/mL).*—Accurately weigh 50 mg USP folic acid that has been dried to constant weight and suspend in ca 30 mL 0.01 N NaOH. Add 300 mL H_2O and adjust to pH 7–8 with HCl. Dilute to 500 mL with H_2O and store in refrigerator under toluene.

(b) *Intermediate solution (1 µg/mL).*—Dilute 5 mL stock solution with H_2O and adjust to pH 7–8 with HCl. Dilute to 500 mL with additional H_2O. Store in refrigerator under toluene.

(c) *Working solution.*—Prepare day of assay. Take 1 mL of intermediate solution (1 µg/mL) and follow procedure **E(c)***(1)*, the concentration of standard solution is now 10 ng/mL. Remove 10 mL of standard solution and dilute to 500 mL with H_2O including 10 mL pH 6.8 buffer [**E(c)***(2)*] (the final volume contains a minimum of 2 mL buffer for every 100 mL final volume). This is the primary standard.

In addition, run 2 secondary standards, one higher (usually 0.3 ng/mL) and one lower (usually 0.12 ng/mL) than the primary standard, in case the working solution is either too high or too low for the primary standard to be used.

(*Note:* If last dilution step requires >10 mL aliquot, match the amount taken with pH 6.8 buffer before final volume is obtained.)

D. Inoculum

(a) *Test organism.*—*Lactobacillus casei* (ATCC 7469) maintained through weekly transfers on agar maintenance medium incubated for 16–21 h at 37°C. Store in refrigerator.

(b) *Agar maintenance medium.*—Use **300J(b)**.

(c) *Liquid culture medium.*—Prepare solubilized liver solution (1 mg/mL) by suspending 0.1 g Difco liver (No. 0133-01) in 100 mL H_2O, hold mixture for 1 h at 50°C, filter, and store filtrate under toluene in refrigerator. Dilute a measured volume of basal medium with an equal volume of H_2O containing 0.8 ng folic acid and 20 µg solubilized liver per mililiter. Add 10 mL portions of diluted medium to 20×150 mm screw cap test tubes, autoclave 15 min at 121–123°C, and cool tubes rapidly. Store in refrigerator. Difco bacto micro inoculum broth (No. 0320-02) has been found satisfactory for liquid culture medium.

(**d**) *Working inoculum.*—Transfer the day before assay is to be run. Transfer cells from *L. casei* stock culture to two sterile tubes containing the liquid culture medium, keeping all transfers as sterile as possible. Incubate in constant temperature bath 16 h at 37°C. Under aseptic conditions, centrifuge culture and decant supernatant. Wash cells with 10 mL sterile saline [**300G(c)**]. Repeat procedure two more times. After the third washing with sterile saline, resuspend cells. Add 5 drops to 10 mL sterile saline. Cell suspension is the inoculum.

E. Assay Solution

(Use distilled H_2O unless otherwise indicated.)

(**a**) Prepare 1.42% Na_2HPO_4 buffer solution using glass distilled H_2O or highest purity available. Deionized H_2O or reverse osmosis device H_2O may contain traces of organic contaminants that inhibit enzymes. Dissolve 1.42 g Na_2HPO_4 and 1.0 g of ascorbic acid per 100 mL of buffer needed. Adjust to pH 7.8 with 4 N NaOH.

[*Note:* 30 mL of buffer is needed for each sample and standard plus an additional 4 mL per sample and standard for the conjugase solution in **E(b)**.]

(**b**) *Conjugase (chicken pancreas) solution (5 mg/mL).*—Weigh out sufficient amount of Difco chicken pancreas (No. 0459-12) (4 mL of extract is needed for each sample and standard). Add to appropriate mL of buffer [**E(a)**] to obtain 5 mg/mL conjugase solution. Stir solution vigorously for 10 min. Transfer to two 20×150 mm test tubes and centrifuge for 10 min at ca 2000 rpm. Decant supernatant through glass wool pledget into beaker and cover with parafilm. Store in refrigerator.

(**c**) *Samples and standard.—(1) pH 4.5.*—1 μg/mL standard is run with samples. Accurately measure an amount of sample equal to 0.25–1.0 g dry solids containing ca 1 μg folic acid. (If sample is low in folic acid, do not take >1.0 g. Increase the second serial dilution to compensate.) Transfer sample to 125 mL Erlenmeyer flask containing 10 mL buffer [**E(a)**] and mix thoroughly. Add an additional 10 mL of buffer and 0.1 mL of octanol as an antifoaming agent. Follow the same procedure for the standard (1 mL of standard). Cover flask with 50 mL beaker, autoclave 15 min at 121–123°C, cool, and add an additional 10 mL of buffer. Add 4 mL of conjugase solution [**E(b)**] to each flask. Add 1 mL toluene to each flask (optional), cover, and incubate 16 h or overnight at 37°C. Inactivate enzyme by autoclaving sample and standard 3 min at 100°C and cool. Adjust sample and standard to pH 4.5 with HCl, dilute to final volume of 100 mL with H_2O, and filter. Follow **C(c)** with standard.

(2) pH 6.8.—Prepare 1.42% Na_2HPO_4 solution as in **E(a)** using distilled H_2O and adjust to pH 6.8 with 4 N NaOH. *This is pH 6.8 buffer.* Take known aliquot of clear filtrate and dilute to final volume with folic acid concentration of 0.2 ng/mL (final volume contains ≥2 mL of pH 6.8 buffer for each 100 mL of final volume).

(*Note:* If last dilution step requires >10 mL aliquot, match the amount taken with pH 6.8 buffer before final volume is obtained.)

F. Assay

Using standard solution [**332C(c)**], assay solution (**332E**), basal media (**332B**), and inoculum [**332D(d)**], proceed as in **300J** and **K**.

References

(1) *Official Methods of Analysis* (1995) 16th Ed., AOAC INTERNATIONAL, Gaithersburg, MD, methods **944.12H** (modified), **960.46D**, **G**, and **H**.

(2) Ford, J. E., Salter, D. N., and Scott, K. J. (1969) *J. Dairy Res.* **36**, 435–446.

(3) Flynn, L. M. (1965) *J. Assoc. Off. Anal. Chem.* **48**, 1230–1235.

(4) Toepfer, E. W., Zook, E. G., Orr, M. L., and Richardson, L. R. (1951) *U.S. Department of Agriculture Handbook No. 29*, U.S. Department of Agriculture, Washington DC.

333   Total Folates
Tri-Enzyme

This procedure is a modification of AOAC Official Method **960.46H** (1). The extraction procedure is from Ford (2) and the assay procedure is from Flynn (3) and USDA Handbook No. 29 (4). The tri-enzyme procedure is a modification of Martin (5).

A. Principle

Folic acid (pteroylglutamic acid) occurs naturally in foods as bound glutamic acid residues of varying chain lengths. Most of these bound forms cannot be utilized by the organism and therefore must be liberated prior to microbiological analysis. Folates are hydrolyzed to folic acid by using natural sources of conjugase enzyme such as chicken pancreas. Protein and carbohydrates can bind and absorb folates. Treating high-protein samples with protease and treating high-carbohydrate samples with alpha-amylase increases measured folates. Therefore, a combination of the three enzymes is used to increase the release of folic acid, which is then assayed using the conventional microbiological method.

B. Basal Medium

(**a**) *Folic acid-free double-strength basal medium.*—(Stock basal medium solutions, *see* **300F**).

mL of basal medium to prepare	250	500	1000
(Add in order listed) (mL)			
Adenine–guanine–uracil solution [**300F(b)**]	2.5	5	10
Xanthine solution [**300F(r)**]	5	10	20
Asparagine solution [**300F(c)**]	15	30	60
Vitamin solution III [**300F(n)**]	50	100	200
Salt solution B [**300F(j)**]	5	10	20
PABA-vitamin B_6 solution [**300F(u)**]	2.5	5	10
Add ca 100 mL H_2O and the following solids (g):			
Vitamin-free casein, hydrolyzed	2.5	5	10
Dextrose, anhydrous	10	20	40

mL of basal medium to prepare	250	500	1000
(Add in order listed)			
Potassium phosphate dibasic	0.25	0.5	1
Potassium phosphate monobasic	0.25	0.5	1
Sodium acetate·3H$_2$O	16.6	33.2	66.4
Glutathione	.00125	.0025	.005

Dissolve the following solids (g) in dilute HCl and add to above solution:

L-cysteine·HCl	0.125	0.25	0.5
D,L-tryptophan	0.05	0.1	0.2

Mix, adjust to pH 6.8 with NaOH and add the following (mL):
Dilute to volume with H$_2$O.

(b) The basal medium can also be prepared by using commercially available Difco folic acid casei medium (No. 0822-15) as directed.

C. Standard Solutions

(Use dark bottles.)

(a) *Stock solution (100 µg/mL).*— Accurately weigh 50 mg USP folic acid that has been dried to constant weight and suspend in ca 30 mL 0.01 *N* NaOH. Add 300 mL H$_2$O and adjust to pH 7–8 with HCl. Dilute to 500 mL with H$_2$O and store in refrigerator under toluene.

(b) *Intermediate solution (1 µg/mL).*—Dilute 5 mL stock solution with H$_2$O and adjust to pH 7–8 with HCl. Dilute to 500 mL with additional H$_2$O and store in refrigerator under toluene.

(c) *Working solution.*—(Prepare day of assay.) Remove 1 mL of intermediate solution (1 µg/mL) and follow procedure **E(c)**(*1*). The concentration of the standard solution is now 10 ng/mL.

Remove 10 mL standard solution and dilute to 500 mL with H$_2$O and 10 mL pH 6.8 buffer [**E(c)**(*2*)] (final volume contains ≥2 mL buffer for every 100 mL of final volume). This is the primary standard.

In addition, run two secondary standards, one higher (usually 0.3 ng/mL) and one lower (usually 0.12 ng/mL) than the primary standard, following the same procedure above in case the sample is either too high or too low for the primary standard to be used.

(*Note*: If last dilution step requires >10 mL aliquot, match the amount taken with pH 6.8 buffer before final volume is obtained.)

D. Inoculum

(a) *Test organism.*—*Lactobacillus casei* (ATCC 7469) maintained through weekly transfers on agar maintenance medium incubated for 16–21 h at 37°C. Store in refrigerator.

(**b**) *Agar maintenance medium.*—Use **300G(b)**.

(**c**) *Liquid culture medium.*—Prepare solubilized liver solution (1 mg/mL) by suspending 0.1 g Difco liver (No. 0133-01) in 100 mL H_2O, hold mixture for 1 h at 50°C and filter. Store filtrate under toluene in refrigerator.

Dilute a measured volume of basal medium with an equal volume of H_2O containing 0.8 ng folic acid and 20 μg solubilized liver per mililiter. Add 10 mL portions of diluted medium to 20×150 mm screw cap test tubes, autoclave 15 min 121–123°C, and cool tubes rapidly. Store in refrigerator. [Difco bacto micro inoculum broth (No.0320-02) has also been found satisfactory for liquid culture medium.]

(**d**) *Working inoculum.*—Transfer the day before assay is being run. Transfer cells from stock culture of *L. casei* to two sterile tubes containing liquid culture medium. Keep all transfers as sterile as possible. Incubate in a constant temperature bath for 16 h at 37°C. Under aseptic conditions, centrifuge the culture and decant the supernatant. Wash the cells with 10 mL sterile saline [**300G(c)**]. Repeat procedure two more times. After third washing with sterile saline, resuspend the cells. Add 5 drops to 10 mL of sterile saline. Cell suspension obtained is the inoculum.

E. Assay Solution

(Use distilled H_2O unless otherwise indicated.)

(**a**) Prepare 1.42% Na_2HPO_4 buffer solution using glass-distilled H_2O or highest purity available (deionized H_2O or reverse osmosis device H_2O may contain traces of organic contaminants that inhibit enzymes). Dissolve 1.42 g Na_2HPO_4 and 1.0 g ascorbic acid per 100 mL buffer needed. Adjust pH to 7.8 with 4 N NaOH.

[*Note*: 30 mL of buffer is needed for each sample and standard plus an additional 4 mL per sample and standard for the conjugase solution in **E(b)**.]

(**b**) *Enzyme preparations.*—*(1) Conjugase (chicken pancreas) solution (5 mg/mL).* Weigh out sufficient amount of Difco chicken pancreas (No. 0459-12) (4 mL of extract is needed for each sample and standard). Add appropriate volume of buffer [**E(a)**] to obtain 5 mg/mL extracting solution and vigorously stir solution for 10 min. Transfer to two 20×150 mm test tubes and centrifuge for 10 min at ca 2000 rpm. Decant supernatant through glass wool pledget into beaker, cover with parafilm, and store in refrigerator. *(2) Alpha-amylase solution (20 mg/mL).* Dissolve 0.5 g alpha-amylase (Sigma Chemical Co., St. Louis, MO, No. A-0273) in 25 mL glass-distilled H_2O. Store in refrigerator. *(3) Protease (pronase E) solution (2 mg/mL).* Dissolve 0.05 g pronase (Sigma Chemical Co., No. P-5147) in 25 mL of glass-distilled H_2O. Filter through glass wool pledget if necessary and store in refrigerator.

(**c**) *Samples and standard.*—1 (μg/mL standard is run with samples.) *(1)* Accurately measure an amount of sample equal to 0.25–1.0 g dry solids containing ca 1 μg folic acid. (If sample is low in folic acid, do not take >1.0 g. To compensate, increase the second serial dilution.) Transfer sample to 125 mL Erlenmeyer flask containing 10 mL buffer [**332D(a)**] and mix thoroughly. Add an additional 10 mL of buffer and 0.1 mL of octanol as antifoaming agent. (Follow the same procedure for 1 mL of standard.)

Cover flask with 50 mL beaker, autoclave 15 min at 121–123°C, and cool. Add an additional 10 mL of buffer. Add 4 mL of conjugase solution [**E(b)**(*1*)] and 1 mL alpha-amylase [**E(b)**(*2*)] to each flask. Cover and incubate 4 h at 37°C. After 4 h, add 1 mL protease [**E(b)**(*3*)] and incubate overnight at 37°C. Inactivate enzyme by autoclaving 3 min at 100°C and cool. Adjust samples and standards to pH 4.5 with HCl, dilute to 100 mL with H_2O, and filter.

(*2*) Prepare 1.42% Na_2HPO_4 solution as in **E(a)** using distilled H_2O and adjust to pH 6.8 with 4 *N* NaOH. *This is pH 6.8 buffer.* Remove a known aliquot of clear filtrate and dilute to final volume with a folic acid concentration of 0.2 ng/mL containing ≥2 mL pH 6.8 buffer for each 100 mL final volume).

(*Note*: If last dilution step requires >10 mL aliquot, match the amount taken with pH 6.8 buffer before final volume is obtained.)

F. Assay

Using standard solution [**330.3B(c)**], assay solution (**330.3D**), basal media (**330.3A**), and inoculum [**330.3C(d)**], proceed as in **300E** and **F**.

References

(1) *Official Methods of Analysis* (1995) 16th Ed., AOAC INTERNATIONAL, Gaithersburg, MD, methods **944.12H** (modified), **960.46D, G**, and **H**.

(2) Ford, J. E., Salter, D. N., and Scott, K. J. (1969) *J. Dairy Res.* **36**, 435–446.

(3) Flynn, L. M. (1965) *J. Assoc. Off. Anal. Chem.* **48**, 1230–1235.

(4) Martin, J. I., Landen, W. O., Soliman, A. M., and Eitenmiller, R. R. (1990) *J. Assoc. Off. Anal. Chem.* **73**(5), 805–808.

(5) Toepfer, E. W., Zook, E. G., Orr, M. L., and Richardson, L. R. (1951) *U.S. Department of Agriculture Handbook No. 29*, U.S. Department of Agriculture, Washington DC.

340 Niacin
AOAC

A. Basal Medium

(a) *Niacin-free double-strength basal medium.*—(Stock basal medium solutions, *see* **300F**).

mL of basal medium to prepare *(Add in order listed)* (mL)	250	500	1000
Cystine–tryptophan solution [**300F(e)**]	25	50	100
Adenine–guanine–uracil solution [**300F(b)**]	5	10	20
Vitamin solution IV [**300F(o)**]	5	10	20
Vitamin solution V [**300F(p)**]	5	10	20
Salt solution A [**300F(i)**]	5	10	20
Salt solution B [**300F(j)**]	5	10	20
Add ca 100 mL H_2O and the following solids (g):			
Vitamin-free casein, hydrolyzed	2.5	5	10
Dextrose, anhydrous	10	20	40
Sodium acetate, anhydrous	5	10	20

Swirl to dissolve solids, adjust to pH 6.8 with HCL or NaOH, and dilute to volume with H_2O.

(b) The basal medium can also be prepared by using commercially available Difco niacin assay medium (No. 0322-15).

B. Standard solutions

(a) *Stock solution (100 μg/mL).*—Weigh 50 mg niacin that has been dried to constant weight. Dissolve in 25% ethanol and dilute to 500 mL with additional 25% ethanol.

(b) *Intermediate solution (1 μg/mL).*—Dilute 5 mL of stock solution to 500 mL with 25% ethanol.

(c) *Working solution.*—Prepare day of assay. Dilute 5 mL intermediate solution to 500 mL with H_2O (0.01 µg/mL). This is the primary standard with an equivalent assay rate to the samples.

Also run two secondary standards, one higher (usually 0.02 µg/mL) and one lower (usually 0.08 µg/mL) in case the sample is either too high or too low for the primary standard to be used.

C. Inoculum

(a) *Test organism.*—*Lactobacillus plantarum* (ATCC 8014) maintained through weekly transfers on agar maintenance medium incubated for 16–21 h at 37°C. Store in refrigerator.

(b) *Agar maintenance medium.*—Culture transferred weekly. Use **300G(b)**.

(c) *Liquid culture medium.*—Dilute measured volume of basal medium with an equal volume of H_2O containing 0.2 µg niacin per mililiter. Add 10 mL portions of diluted medium to 20×150 mm screw top test tubes, autoclave 15 min at 121–123°C, and cool tubes as fast as possible. Store in refrigerator.

(d) *Working inoculum.*—Transfer the day before assay is being run. Transfer cells from *L. plantarum* stock culture to two sterile tubes containing liquid culture medium, keeping all transfers as sterile as possible. Incubate in a constant temperature bath for 16–21 h at 37°C. Under aseptic conditions, centrifuge the culture and decant the supernatant. Wash the cells with 10 mL sterile saline [**300G(c)**]. Repeat procedure two more times. After the third washing with sterile saline, resuspend the cells. Cell suspension obtained is the inoculum.

D. Assay Solution

(a) Accurately weight a sample containing 10–30 µg niacin per gram and transfer to a 250 mL beaker.

(b) Add known volume of 1 N H_2SO_4 equal in mililiters to $\geq$10 times the dry weight of the sample in g. Disperse sample evenly in the liquid.

(c) Autoclave mixture 30 min at 121–123°C and cool. If lumping occurs, agitate mixture until particles are evenly dispersed.

(d) Adjust the mixture with vigorous stirring to pH 6.0–6.5 with NaOH. Add dilute HCl until no further precipitation occurs (ca pH 4.5, the isoelectric point for protein).

(e) Dilute mixture to first serial dilution with H_2O and filter.

(f) Remove known aliquot of the clear filtrate, check for complete precipitation with dilute HCl, and proceed as follows: *(1)* If no further precipitation occurs, adjust the clear filtrate to pH 6.8 with NaOH and dilute with H_2O to the final measured volume (assay rate is usually 0.01 µg/mL). *(2)* If further precipitation occurs, adjust the mixture again to a point of maximum precipitation, dilute with H_2O to a known volume, filter, take known aliquot of the clear filtrate, adjust pH to 6.8 with NaOH, and dilute with H_2O to the final measured volume (assay rate is usually 0.01 µg/mL).

E. Assay

Using standard solution [**340B(c)**], assay solution (**340D**), basal media (**340A**), and inoculum [**340C(d)**], proceed as in **300E** and **F**.

Reference

(1) *Official Methods of Analysis* (1995) 16th Ed., AOAC INTERNATIONAL, Gaithersburg, MD, method **944.13I**.

350   Panthenol

A. Basal Medium

(a) *Panthenol-free double-strength basal medium.*—(Stock basal medium solutions, *see* **300F**.)

mL of basal medium to prepare *(Add in order listed)* (mL)	250	500	1000
Cystine solution [**300F(d)**]	18.75	37.5	75
Trytophan solution [**300F(k)**]	25	50	100
Adenine-guanine-uracil solution [**300F(b)**]	2.5	5	10
Salt solution A [**300F(m)**]	5	10	20
Norit A treated Casitone solution [**300F(v)**]	25	50	100
Liver extract solution [**300F(w)**]	1.25	2.5	5
Add ca 100 mL H_2O and the following (g or mL):			
Vitamin-free casein, hydrolyzed	0.5	1	2
Dextrose, anhydrous	3.7	7.5	15
Sodium citrate	0.5	1	2
Glycerol	8.25	16.5	33
Amino acid and vitamin mixture [**300F(x)**]	.005	.1	.2
Ethanol	0.05	0.1	0.2
Lactic acid	0.15	0.3	0.6
Polysorbate 80 (Tween 80)	0.25	0.5	1

Swirl to dissolve solids, adjust pH to 6.0 with HCl and dilute to volume with H_2O.

(b) The basal medium can also be prepared by using commercially available Difco Panthenol assay medium (No. 0994-15) and Difco Panthenol supplement (No. 0212-64). For each liter of basal medium needed, use 33 g Panthenol assay medium and 100 mL Panthenol supplement.

B. Standard Solutions

(**a**) *Stock solution.*—(Equivalent to 50 µg pantoic acid per mililiter.) Accurately weigh 69.2 mg Panthenol. Dissolve in H_2O, adjust to pH 6.0, and dilute to 1 L.

(**b**) *Working solution.*—Prepare day of assay. Remove 3, 4, and 5 mL portions of stock solution and add H_2O for 10 mL total volume in separate 125 mL Erlenmeyer flasks. Add 10 mL 0.1 *N* NaOH to each flask, autoclave 30 min at 121–123°C, and cool. Adjust pH to 6.0 with 0.2 *N* HCl (never go below pH 5.0, otherwise lactone forms). Dilute to 500 mL with H_2O. This will produce one primary standard (0.4 µg/mL) and two secondary standards (0.3 and 0.5 µg/mL).

C. Inoculum

(**a**) *Test organism.*—*Acetobacter suboxydans* (ATCC 621H) maintained through weekly transfers on agar maintenance medium incubated for 16–48 h at 30°C. Store in refrigerator.

(**b**) *Agar maintenance medium.*—Dissolve 5 g mannitol, 0.5 g Difco yeast extract (No. 0127-01), 0.5 g Difco peptone (No. 0118-01), and 1.5 g Difco agar (No. 0140-01) per 100 mL of H_2O. Heat medium to a boil. Place 7 mL of hot agar into 20 × 150 mm test tubes, plug with cotton, and cap. Autoclave 15 min at 121–123°C, tilt hot agar tubes to form slants, and cool in position. Store in refrigerator.

(**c**) *Liquid culture medium.*—Dilute measured volume of basal medium (with the Norit A treated casitone omitted and the hydrolyzed casein increased 10-fold) with an equal volume of H_2O containing 0.02 µg pantoic acid per mililiter. Add 10 mL portions of diluted medium to 20 × 150 mm screw cap test tubes each containing two 4–5 mm glass beads. Autoclave 10 min at 100°C, cool tubes rapidly, and store in refrigerator.

(**d**) *Working inoculum.*—(Transfer the day before assay is run.) Transfer cells from *A. suboxydans* stock culture to two sterile tubes containing liquid culture medium, keeping all transfers as sterile as possible. Incubate for 16–21 h on rotary shaker in a constant temperature bath or room at 30°C. Under aseptic condition, centrifuge the culture and decant supernatant. Wash cells with 10 mL sterile saline [**300G(c)**]. Repeat the procedure two more times. Resuspend cells and add 5 mL of cell suspension to 10 mL sterile saline. Cell suspension obtained is the inoculum.

D. Assay Solution

(**a**) *Extraction of sample.*—*(1)* Samples containing no pantoic acid, pantoyl lactone, or pantothenates can be extracted in H_2O and diluted to 20 µg Panthenol per mililiter. Solids are placed in blender, diluted to same potency, and filtered. These extracts may be hydrolyzed without resin purification treatment. Proceed as in **D(c)**. *(2)* Samples containing significant amounts of pantoic acid, pantoyl lactone, or pantothenic acid are extracted and diluted with 0.12 *N* sodium citrate buffer (pH 5) to a Panthenol concentration of 100 µg/mL (levels as low as 12 µg/mL may be used), and filtered. These extracts must go through resin purification treatment. Proceed as in **D(b)**. *(3)* Samples containing ethanol must be evaporated to dryness on steam bath under a stream of air to remove alcohol. Proceed as in step **D(a)***(1)* or **(a)***(2)*.

(**b**) *Resin purification treatment.*—To remove pantothenates, use column 14–15 × 200 mm with fritted end and petcock, pack to 4 cm with amberlite IRA-400 (OH) 100–200 mesh, and place glass wool pledget on top of packing. (For more difficult samples, *see* below.)

Wash bed with successive portions 2 *N* HCl totaling 50 mL and then with 25 mL H$_2$O. Add 50 mL 10% NaOH solution to column. Remove excess OH by successive washings of H$_2$O totaling 50 mL. (Prepare fresh bed for each sample.) As H$_2$O level reaches resin, place a 50 mL volumetric flask under column tip and place 10 mL sample onto column. When no fluid remains above resin bed, add 40 mL H$_2$O. Collect and dilute to volume and proceed as in (**c**).

For more difficult samples, use a column 14–15 × 20 mm with fritted end and petcock with the following combination of resins: Adsorbent: 3 cm packed florosil 60–80 mesh (5 cm for large amounts of riboflavin). Anion exchanger: 3 cm packed amberlite IRA-400 (Cl or OH) 100–200 mesh. Cation exchanger: 3 cm packed Dowex-50 100–200 mesh. Thoroughly wash column with ca 20 mL H$_2$O. As H$_2$O level reaches resin, place a 50 mL volumetric flask under column tip and place 10 mL sample onto column. When no fluid remains above resin bed, add 40 mL H$_2$O. Collect and dilute to volume and proceed as in (**c**).

(**c**) *Sample hydrolysis.*—(Both standard and sample are hydrolyzed.) Add 10 mL aliquot (20 µg/mL) to a 125 mL Erlenmeyer flask. Add 10 mL 0.1 *N* NaOH to each flask, autoclave 30 min at 121–123°C, and cool. Adjust pH to 6.0 with 0.2 *N* HCl (never go below pH 5.0 otherwise lactone forms). Dilute to 500 mL with H$_2$O.

E. Assay

Using standard solution [**350B(b)**], assay solution [**350D(c)**], basal media (**350A**), and inoculum [**350C(d)**], proceed as in **300J** and **K**.

[*Note*: Multiplication factors are as follows: *(1)* Multiply the potency found by 1.384 to convert pantoic acid to Panthenol. *(2)* Multiply the potency found by 1.478 to convert pantoic acid to pantothenic acid.]

References

(1) *Official Methods of Analysis* (1995) 16th Ed., AOAC INTERNATIONAL, Gaithersburg, MD, method **960.46D**, **G**, and **H**.

(2) Weiss, M. S., Sonnenfeld, I., De Ritter, E., and Rubin, S. H. (1951) *Anal. Chem.* **23**, 1687–1689.

(3) De Ritter, E. and Rubin, S. H. (1949) *Anal. Chem.* **21**, 823–827.

(4) Sarett, H. P. and Cheldelin, V. H. (1945) *J. Biol. Chem.* **159**, 311–319.

360   Pantothenic Acid
AOAC

(Applicable to products containing free pantothenic acid.)

A. Basal Medium

(**a**) *Pantothenic acid-free double-strength basal medium.*—(Stock basal medium solutions, *see* **300F**.)

mL of basal medium to prepare *(Add in order listed)* (mL)	250	500	1000
Cystine–tryptophan solution [**300F(e)**]	25	50	100
Adenine–guanine–uracil solution [**300F(b)**]	5	10	20
Vitamin solution IV [**300F(o)**]	5	10	20
Vitamin solution VI [**300F(q)**]	5	10	20
Salt solution A [**300F(i)**]	5	10	20
Salt solution B [**300F(j)**]	5	10	20
Add ca 100 mL H$_2$O and the following (g or mL):			
Vitamin-free casein	2.5	5	10
Dextrose, anhydrous	10	20	40
Sodium acetate·3H$_2$O	8.3	16.6	33.2
Polysorbate 80 solution [**300F(h)**]	0.25	0.5	1

Swirl to dissolve solids, adjust pH to 6.8 with NaOH and dilute to volume with H$_2$O.

(**b**) The basal medium can also be prepared by using commercially available Difco pantothenate medium AOAC USP (No. 0816-15).

B. Standard Solutions

(**a**) *Stock solution (40 μg/mL).*—Accurately weigh 43.47 mg USP calcium pantothenate that has been dried to constant weight. Dissolve in 500 mL H$_2$O, add 10 mL 0.2 *N* acetic acid and 100 mL 0.2 *N* sodium acetate, and dilute to 1 L with additional H$_2$O. Store in refrigerator under toluene.

(b) *Intermediate solution (1 μg/mL).*—Dilute 25 mL stock solution with ca 500 mL H_2O. Add 10 mL 0.2 N acetic acid and 100 mL 0.2 N sodium acetate. Dilute to 1 L with additional H_2O. Store in refrigerator under toluene.

(c) *Working solution.*—(Prepare day of assay.) Dilute 5 mL intermediate solution to 500 mL with H_2O (0.01 μg/mL). This is the primary standard with an equivalent assay rate to the samples.

Also run two secondary standards, one higher (usually 0.014 μg/mL) and one lower (usually 0.008 μg/mL) than the primary standard in case the sample is either too high or too low for the primary standard to be used.

C. Inoculum

(a) *Test organism.*—*Lactobacillus plantarum* (ATCC 8014) maintained through weekly transfers on agar maintenance medium incubated for 16–21 h at 37°C. Store in refrigerator.

(b) *Agar maintenance medium.*—Culture transferred weekly. Use **300G(b)**.

(c) *Liquid culture medium.*—Dilute a measured volume of basal medium with an equal volume of H_2O containing 0.04 μg pantothenic acid per mililiter. Add 10 mL portions of diluted medium to 20 × 150 mm screw cap test tubes, autoclave 15 min at 121–123°C, and cool tubes as fast as possible. Store in refrigerator.

(d) *Working inoculum.*—(Transfer the day before assay is being run.) Transfer cells from *L. plantarum* stock culture to two sterile tubes containing liquid culture medium, keeping all transfers as sterile as possible. Incubate for 16–21 h in a constant temperature bath at 37°C. Under aseptic conditions, centrifuge the culture and decant the supernatant. Wash the cells with 10 mL sterile saline [**300G(c)**]. Repeat the procedure two more times. After the third washing with sterile saline, resuspend the cells. Cell suspension obtained is the inoculum.

D. Assay Solution

(Keep solution below pH 7.0 to prevent loss or destruction of the vitamin.)

(a) Place measured amount of sample in 250 mL beaker.

(b) Add volume of H_2O equal in mililiters to 10 times the dry weight of sample. Resulting solution must not contain >5 mg of pantothenic acid per mililiter.

(c) Adjust mixture to pH 5.6–5.7 with acetic acid or sodium acetate. If the sample is not readily soluble, comminute so it is evenly dispersed in liquid. Agitate vigorously and wash down sides of beaker with solution containing 10 mL 0.2 N acetic acid and 100 mL 0.2 N sodium acetate per liter.

(d) Autoclave 5–7 min at 121–123°C and cool. If lumping occurs, agitate mixture until particles are evenly dispersed.

(e) Stirring vigorously, adjust mixture to pH 6.0–6.5 with NaOH. Add dilute HCl to pH 4.5 to precipitate the protein.

(f) Dilute mixture to first serial dilution with H_2O and filter.

(**g**) Remove known aliquot of the clear filtrate, check for complete precipitation with dilute HCl, and proceed as follows: *(1)* If no further precipitation occurs, adjust the clear filtrate to pH 6.8 with NaOH and dilute to the final measured volumen with H_2O (assay rate is ca 0.01 µg/mL). *(2)* If further precipitation occurs, adjust the mixture again to a point of maximum precipitation, dilute with H_2O to a known volume, filter, take known aliquot of the clear filtrate, and follow (**g**)*(1)*.

E. Assay

Using standard solution [**360B(c)**], assay solution (**360D**), basal media (**360A**), and inoculum [**360C(d)**], proceed as in **300J** and **K**.

[*Note:* Multiplication factors are as follows: *(1)* Multiply potency found by 1.087 if expressed as calcium pantothenate. *(2)* Multiply potency found by 1.100 if expressed as sodium pantothenate.]

Reference

(1) *Official Methods of Analysis* (1995) 16th Ed., AOAC INTERNATIONAL, Gaithersburg, MD, method **945.75I**.

361 Total Pantothenates

AOAC Infant Formula

This procedure is a modification of AOAC® Official Method **945.74** (1). The extraction and assay solution procedures are taken from USDA Handbook No. 97 (2).

A. Principle

Samples of foods are subjected to simultaneous attack by the enzymes intestinal alkaline phosphatase and avian liver peptidase. The intestinal phosphatase removes the triphosphoadenine from co-enzyme A and the avian liver enzyme breaks the peptide linkage. Bound pantothenic acids are released in the free form and are then assayed microbiologically.

B. Basal Medium

(**a**) *Preparation of pantothenic acid-free double-strength basal medium.*—(Stock basal medium solutions, see **300F**.)

mL of basal medium to prepare	250	500	1000
(Add in order listed) (mL)			
Cystine–tryptophan solution [**300F(e)**]	25	50	100
Adenine–guanine–uracil solution [**300F(b)**]	5	10	20
Vitamin solution IV [**300F(o)**]	5	10	20
Vitamin solution VI [**300F(q)**]	5	10	20
Salt solution A [**300F(i)**]	5	10	20
Salt solution B [**300F(j)**]	5	10	20
Add ca 100 mL H_2O and the following (g or mL):			
Vitamin-free casein	2.5	5	10
Dextrose, anhydrous	10	20	40
Sodium acetate·$3H_2O$	8.3	16.6	33.2
Polysorbate 80 solution [**300F(h)**]	0.25	0.5	1

Swirl to dissolve solids, adjust to pH 6.8 with NaOH, and dilute to volume with H_2O.

(**b**) The basal medium can also be prepared by using commercially available Difco pantothenate medium AOAC USP (No. 0816-15).

41

C. Standard Solutions

(**a**) *Stock solution (40 μg/mL).*—Accurately weigh 43.47 mg USP calcium pantothenate that has been dried to constant weight. Dissolve in 500 mL H_2O, add 10 mL 0.2 *N* acetic acid and 100 mL 0.2 *N* sodium acetate, and dilute to 1 L with additional H_2O. Store in refrigerator under toluene.

(**b**) *Intermediate solution (1 μg/mL).*—Dilute 25 mL stock solution with ca 500 mL H_2O. Add 10 mL 0.2 *N* acetic acid and 100 mL 0.2 *N* sodium acetate. Dilute to 1 L with additional H_2O. Store in refrigerator under toluene.

(**c**) *Working solution.*—(Prepare day of assay.) *(1)* Remove 5 mL of stock solution (40 μg/mL) and follow procedure in **E(b)**(*3*). *(2)* After following the procedure in sample preparation through step **E(b)**(*8*), the concentration of pantothenic acid is 0.02 μg/mL. Remove 50 mL of clear aliquot from the combined standard portions, adjust to pH 6.8, and dilute to 100 mL containing 0.01 μg/mL. This is the primary standard and has the same assay concentration as the samples. *(3)* In addition, run two secondary standards, one higher (usually 0.008 μg/mL) and one lower (usually 0.014 μg/mL). Remove a clear aliquot from the remaining standard portion (40 mL for 0.008 μg/mL and 70 mL for 0.014 μg/mL), adjust pH to 6.8, and dilute to 100 mL to prepare the secondary standards. This procedure brackets the samples in case they are higher or lower in concentration than the primary standard. (*Note:* Standards must be hand-pipeted due to small volumes.)

D. Inoculum

(**a**) *Test organism.*—Lactobacillus plantarum (ATCC 8014) maintained through weekly transfers on agar maintenance medium incubated for 16–21 h at 37°C. Store in refrigerator.

(**b**) *Agar maintenance medium.*—(Culture transferred weekly.) Use **300C(b)**.

(**c**) *Liquid culture medium.*—Dilute measured volume of basal medium with equal volume of H_2O containing 0.04 μg pantothenic acid per mililiter. Add 10 mL portions of diluted medium to 20 × 150 mm screw cap test tubes, autoclave 15 min at 121–123°C, and cool tubes as quickly as possible. Store in refrigerator.

(**d**) *Working inoculum.*—(Transfer the day before assay is run.) Transfer cells from *L. plantarum* stock culture to two sterile tubes containing the liquid culture medium, keeping all transfers as sterile as possible. Incubate for 16–21 h in a constant temperature bath at 37°C. Under aseptic conditions, centrifuge the culture and decant the supernatant. Wash the cells with 10 mL sterile saline [**300C(c)**]. Repeat the procedure twice more. After the third washing with sterile saline, resuspend the cells. Cell suspension obtained is the inoculum.

E. Assay Solution

(a) *Enzyme preparation.*

(1) Phosphatase solution, 2% (20 mg/mL).—Prepare by dissolving intestinal alkaline phosphatase (Sigma Chemical Co., Product No. P-3877) in H_2O. Prepare fresh for each assay and store in refrigerator until needed.

(2) Pigeon liver extract solution, 10%.—Pigeon liver acetone powder (Sigma Chemical Co., No. L-8376).

(a) Equipment.—*(1)* Large mortar and pestle. *(2)* Centrifuge HN S-II (4 hole head). *(3)* 500 mL (0.02 M $KHCO_3$, 1g → 500 mL). *(4)* Twenty-four 50 mL plastic centrifuge tubes with caps. *(5)* Three 500 mL wide-mouth flasks. *(6)* Fifty 16×125 mm test tubes with caps. *(7)* Two 10 mL Mohr pipets. *(8)* Two glass rods. *(9)* Two hard-rubber policemen.

(*Note:* All supplies and equipment above are placed in freezer overnight except for the $KHCO_3$, the Dowex-1, and the centrifuge, which are maintained in a refrigerator or 4°C constant temperature room.)

(b) Procedure.—Quantities for preparation of a triple bath (yields ca 150 mL). Rub 30 g acetone powder liver extract in cold mortar held in an ice-salt bath with ca 150 mL ice-cold 0.02M $KHCO_3$. Quantitatively transfer the suspension to 8 ice-cold centrifuge tubes with the aid of 150 mL $KHCO_3$ for a total of 300 mL $KHCO_3$. Shake the capped centrifuge tubes vigorously and chill in the freezer for 10 min. Centrifuge 5 min at 3000 rpm.

Transfer supernatant through cheesecloth into a cold 500 mL wide-mouth flask. Add 150 g of activated Dowex [**E(a)***(3)*] and shake for 5 min in an ice-salt bath or shaker box in a cold room. Transfer mixture to eight cold centrifuge tubes and centrifuge 5 min at 3000 rpm. Transfer supernatant through new cheesecloth to another cold 500 mL wide-mouth flask. Chill in freezer for 10 min. Supernatant is treated in the same manner as described previously with the remaining 150 g activated Dowex. Place on shaker 5 min, transfer mixture to eight cold centrifuge tubes, and centrifuge 5 min at 3000 rpm. Transfer the supernatant through previously used cheesecloth to another cold 500 mL wide-mouth flask. Chill in freezer for 10 min. Dispense the supernatant into cold 16×125 mm screw cap test tubes in multiples of 0.2 mL (usually 3 mL portions) for use in subsequent assays. **Freeze and store in freezer.** When needed for assay, remove tubes of the frozen enzyme suspension, allow to thaw, then store in refrigerator until needed.

(3) Dowex 1-X8.—(Bio-Rad Laboratories, Inc., Brussels, Belgium.) (Prepare a triple batch). (a) Stir 100 g Dowex 1-X8 (200–400 mesh) for 10 min in 1 L of 1 *N* HCl using mechanical stirrer capable of keeping resin completely suspended. (b) Filter with suction through Whatman No. 50 paper in Büchner funnel. Repeat steps (a) and (b) using another liter of 1 *N* HCl. Repeat steps (a) and (b) using 1 L H_2O. Repeat H_2O wash ten times. Prepare Tris buffer, pH 8.3, [**E(b)***(1)*]. Adjust resin to pH 8.0 by adding Tris buffer, pH 8.3, dropwise. Store slurry in refrigerator and use within two days.

(b) *Sample preparation.*—*(1)* Prepare Tris buffer, pH 8.3, by dissolving 24.2 g 2-amino-2-(hydroxymethyl)-1,3-propanediol in H_2O and dilute to 200 mL. Adjust pH to 8.3 with 30% HCl and filter. Store at room temperature and use within two weeks of preparation date. *(2)* Prepare sodium bicarbonate solution by dissolving 850 mg $NaCO_3$ in H_2O and dilute to 100 mL. Prepare fresh on day of use and store in refrigerator. Place 50 mL of distilled H_2O in 100 mL flask and store in refrigerator. *(3)* Remove a quantity of sample containing 50–100 μg total pantothenates and 5 mL pantothenic acid stock standard solution. With each assay, weigh 1.0 g dry Brewers yeast. Brewers yeast contains ca 100 μg total pantothenates per gram. The yeast is assayed with the samples to determine the efficacy and reproducibility of the enzyme hydrolysis procedure. *(4)* Transfer sample to 125 or 250 mL Erlenmeyer flask containing 10 mL Tris buffer, pH 8.3. (Use Tris buffer, pH 11.3, for very acidic samples low in pantothenates. Larger sample sizes are needed for samples low in pantothenates.) Treat dry Brewers yeast and standard in the same manner as samples. *(5)* Add sufficient H_2O to obtain a good suspension. Add 1.5 mL octanol to prevent foaming and bumping. *(6)* Autoclave 15 min at 121–123°C, cool, and dilute to 200 mL for standard or 100 mL for yeast to a concentration of 1 μg/mL. *(7)* Perform the following operations with 20 × 150 mm test tubes, rack, and all solutions in ice-salt bath. Carefully pipet to deliver solutions to the bottom of the tubes.

Prepare sample tubes containing:
    1.0 mL sample solution (1 μg/mL) [*see* **E(b)**(6)]
    0.1 mL sodium bicarbonate solution [*see* **E(b)**(2)]
    0.4 mL intestinal phosphatase solution [*see* **E(a)**(1)]
    0.2 mL pigeon liver extract solution [*see* **E(a)**(2)]
    0.4 mL distilled H_2O (cold).

Prepare Brewers yeast tubes using the yeast solution in place of the sample solution.

Prepare the pantothenic acid standard tubes using the standard solution [**E(b)**(6)] in place of the sample solution. The pantothenic acid standard is treated with enzymes in the same manner as the samples to eliminate the need for an enzyme blank. Secondary standards should also be prepared. Therefore, four pantothenic acid standard tubes are needed for each assay.

Mix tubes one at a time by carefully rolling tubes between palms. Do not allow mixture to stick onto tube walls. Optional, add 0.1 mL toluene to all tubes. Cover tubes with parafilm and incubate for 4 h or overnight at 37°C.

(8) Adjust enzyme-treated samples, yeast, and standard to pH 4.5 (combine four test tubes before adjusting pH), dilute to measured volume containing 0.02 μg pantothenic acid per mililiter (200 mL for standard and 50 mL for samples and yeast), and filter. For standard, proceed to **C(c)**(2). *(9)* Remove 25 mL aliquot of the clear filtrate, check for complete precipitation with dilute HCl, and proceed as follows: if no further precipi-

tation occurs, adjust the pH of the clear filtrate to 6.8 with NaOH and dilute to 50 mL with H_2O (assay rate is usually 0.01 μg/mL). If further precipitation occurs, adjust the mixture again to a point of maximum precipitation, dilute to a known volume with H_2O, filter, take known aliquot of the clear filtrate, adjust the pH to 6.8 with NaOH, and dilute with H_2O until final volume is obtained (assay rate is 0.01 μg/mL). *(10)* Sample and yeast must be hand pipetd due to small volumes obtained for assay solutions.

F. Assay

Using standard solution [**361C(c)**], assay solution [**361D(b)**], basal media (**361A**), and inoculum [**361C(d)**], proceed as in **300J** and **K**.

[*Note*: Multiplication factors are as follows: (1) Multiply potency found by 1.087 if results are to be expressed as calcium pantothenate. (2) Multiply potency found by 1.100 if results are to be expressed as sodium pantothenate.]

References

(1) *Official Methods of Analysis* (1995) 16th Ed., AOAC INTERNATIONAL, Gaithersburg, MD, methods **945.74I** (modified), **960.46D**, **G**, and **H**.

(2) Zook, E. G., MacArthor, M. J., and Toepfer, E. W. (1956) *U.S. Department of Agriculture Handbook No. 97*, U.S. Department of Agriculture, Washington DC.

370 Riboflavin
(Vitamin B₂) AOAC

A. Basal Medium

(a) *Riboflavin-free double-strength basal medium.*—(Stock basal medium solutions, *see* **300F**.)

mL of basal medium to prepare	250	500	1000
(Add in order listed) (mL)			
Photolyzed peptone solution [**300F(g)**]	50	100	200
Cystine solution [**300F(d)**]	25	50	100
Yeast supplement solution [**300F(s)**]	5	10	20
Salt solution A [**300F(i)**]	5	10	20
Salt solution B [**300F(j)**]	5	10	20

Add an appropriate amount H_2O to dissolve the following (g):			
Dextrose, anhydrous	15	30	60

Swirl until dextrose is dissolved. Adjust pH to 6.8 and dilute to volume with H_2O.

(b) Basal medium can also be prepared using commercially available Difco riboflavin assay medium (No. 0325-15).

B. Standard Solutions

(Use dark bottles.)

(a) *Stock solution (100 µg/mL).*—Weigh 50 mg USP riboflavin dried to constant weight and suspend in ca 300 mL 0.02 *N* (1.2 mL/L) acetic acid and warm on a steam bath, stirring, until solid dissolves. Cool and dilute to 500 mL with 0.02 *N* acetic acid. Store in refrigerator under toluene.

(b) *Intermediate solution (1 µg/mL).*—Dilute 5 mL stock solution to 500 mL with 0.02 *N* acetic acid. Store in refrigerator under toluene.

(c) *Working solutions.*—(Prepare the day of assay.) Dilute 15 mL intermediate solution to 500 mL with H_2O (0.03 μg/mL). This is the primary standard with an equivalent assay rate to the sample.

Also run two secondary standards, one higher (usually 0.04 μg/mL) and one lower (usually 0.02 μg/mL) than the primary standard in case the sample is either too high or too low for the primary standard to be used.

C. Inoculum

(a) *Test organism.*—*Lactobacillus casei* (ATCC 7469) maintained through weekly transfers on agar maintenance medium incubated for 16–21 h at 37°C. Store in refrigerator.

(b) *Agar maintenance medium.*—(Transfer culture weekly.) Use **300G(b)**.

(c) *Liquid culture medium.*—Dilute measured volume basal medium with equal volume of H_2O containing 0.1 μg riboflavin per mililiter. Add 10 mL portions of diluted medium to 20×150 mm screw cap test tubes, autoclave 15 min at 121–123°C, and cool tubes rapidly. Store in refrigerator.

(d) *Working inoculum.*—(Transfer the day before assay is run.) Transfer cells from *L. casei* stock culture to two sterile tubes containing liquid culture medium, keeping all transfers as sterile as possible. Incubate for 16–21 h in a constant temperature bath at 37°C. Under aseptic conditions, centrifuge the culture and decant the supernatant. Wash the cells with 10 mL sterile saline [*see* **300G(c)**]. Repeat procedure two more times. After the third washing with sterile saline, resuspend the cells. Cell suspension obtained is the inoculum.

D. Assay Solution

(Keep solution below pH 7.0 to prevent loss of riboflavin.)

(a) Accurately weigh sample containing 0.5–1.0 μg riboflavin per gram and transfer to 250 mL beaker.

(b) Add 0.1 *N* HCl equal (in mL) to not less than 10 times the dry weight of sample (in grams) (usually 100 mL is a minimum volume). Disperse sample evenly in the liquid.

(c) Agitate sample vigorously, wash down sides of beaker with 0.1 *N* HCl, and cover with watch glass.

(d) Autoclave mixture 30 min at 121–123°C and cool. If lumping occurs, agitate mixture until particles are evenly dispersed.

(e) Adjust the mixture using a pH meter and stirring to pH 6.0–6.5 with 1 *N* NaOH. Add dilute HCl immediately until no further precipitation occurs (usually pH 4.5, the isoelectric point for protein).

(f) Dilute mixture to the first volume in the serial dilutions with H_2O and filter. If the fat content is high, extract a known aliquot of the filtrate three times with 30 mL portions of ether.

(g) Remove known aliquot of clear filtrate, check for complete precipitation with dilute HCl, and if no further precipitation occurs, adjust the clear filtrate to pH 6.8 with NaOH and dilute to the final volume with H_2O (assay rate is ca 0.02 µg/mL).

If further precipitation occurs, adjust the mixture again to point of maximum precipitation, dilute to a known volume with H_2O, filter, and take known aliquot of the clear filtrate. Adjust the clear filtrate to pH 6.8 with NaOH and dilute to the final volume with H_2O (assay rate is ca 0.02 µg/mL).

E. Assay

Using standard solution [**370B(c)**], assay solution (**370D**), basal media (**370A**), and inoculum [**370C(d)**], proceed as in **300J** and **K**.

Reference

(1) *Official Methods of Analysis* (1995) 16th Ed., AOAC INTERNATIONAL, Gaithersburg, MD, method **940.33I**.

380 Thiamine

(Vitamin B₁)

A. Basal Medium

Prepare the basal medium using commercially available Difco thiamine assay medium LV (No. 0808-15) as directed.

B. Standard Solutions

(Use dark bottles.)

(a) *Stock solution (100 µg/mL).*—Accurately weigh 50 mg USP thiamine·HCl dried to constant weight. Dissolve in 20% ethanol, adjusted to pH 3.5–4.3 with HCl, and dilute to 500 mL with acidified ethanol. Store in refrigerator.

(b) *Intermediate solution (10 µg/mL).*—Dilute 100 mL stock solution to 1 L with acidified 20% ethanol. Store in refrigerator.

(c) *Working solutions.*—(Prepare the day of assay.) Dilute 5 mL stock solution to 500 mL with H_2O (1 µg/mL). Remove 10 mL of 1 µg/mL solution and place in 250 mL Erlenmeyer flask with 90 mL of 0.1 N HCl. Autoclave 30 min at 121°C. Cool and adjust pH to 4.5 with 50% sodium acetate. Add 5 mL 10% takadiastase solution, dilute to 200 mL, filter, and incubate 16 h or overnight at 37°C. Remove 40 mL aliquot and adjust pH to 6.0 with 50% sodium acetate. Dilute to 500 mL with H_2O. Standard solution is 4.0 ng/mL. This is the primary standard with an equivalent assay rate to the sample.

Also run two secondary standards, one higher (usually 5.0 ng/mL) and one lower (usually 3.0 ng/mL) in case the sample is either too high or too low for the primary standard to be used.

C. Inoculum

(a) *Test organism.*—*Lactobacillus viridescens* (ATCC 12706) maintained through weekly transfers on agar maintenance medium incubated for 16–21 h at 30°C. Store in refrigerator.

(b) *Agar maintenance medium.*—Use Difco APT agar (No. 0645-15) as directed. Heat to boil. Place 10 mL of hot agar into 20 × 150 mm test tubes, plug with cotton, and cap. Autoclave 15 min at 121–123°C. Cool and store in refrigerator.

(c) *Liquid culture medium.*—Dissolve 23.1 g Difco APT broth (No. 0655-01-7) in H_2O and add 2.5 mL 100 µg/mL thiamine standard. Bring volume up to 500 mL with H_2O. Pipet 10 mL into each test tube and autoclave 15 min at 121°C. Store in refrigerator.

(d) *Working inoculum.*—(Transfer the day before assay is being run.) Transfer cells from *L. viridescens* stock culture to two sterile tubes containing the liquid culture medium, keeping all transfers as sterile as possible. Incubate in a constant temperature bath for 16–221 h at 30°C. Under aseptic conditions, centrifuge the culture and decant the supernatant. Wash the cells with 10 mL sterile saline [**300G(c)**]. Repeat procedure two more times. After the third washing with sterile saline, resuspend the cells. Add 1 mL cell suspension to 10 mL sterile saline. Cell suspension obtained is the inoculum.

D. Assay Solution

(Run standard with sample.)

(a) Accurately weigh sample containing ≥10 µg thiamine and transfer to 250 mL flask.

(b) Add 0.1 *N* HCl (in mL) equivalent to 10 times the dry weight or volume of the sample (ca 100 mL).

(c) Agitate sample vigorously, wash down sides of flask with 0.1 *N* HCl, and cover with small 100 mL beaker.

(d) Autoclave mixture 30 min at 121–123°C and cool. If lumping occurs, agitate mixture until particles are evenly dispersed.

(e) Adjust the mixture using pH meter to pH 4.5 with 50% sodium acetate·$3H_2O$ while stirring, and then dilute to 0.1 µg thiamine per mililiter (usually 100 mL).

(f) Add 5 mL of a 10% takadiastase solution (or equivalent) that has been brought up to volume with H_2O and incubate 16 h or overnight at 37°C.

(g) Readjust pH to 4.5 if necessary. Dilute mixture to measured volume and filter.

(h) Remove known aliquot of the clear filtrate, adjust to pH 6.0 with sodium acetate·$3H_2O$, and dilute to final serial dilution volume with H_2O (assay rate concentration is 4.0 ng thiamine per mililiter).

(i) Inoculate [**C(d)**] and incubate 18 h at 30°C.

E. Assay

Using standard solution [**380B(c)**], assay solution (**380D**), basal media (**380A**), and inoculum [**380C(d)**], proceed as in **300J** and **K**.

References

(1) Defibaugh, P. W., Smith, J. S., and Weeks, C. E. (1977) *J. Assoc. Off. Anal. Chem.* **60**, 522–527.

(2) *Official Methods of Analysis* (1995) 16th Ed., AOAC INTERNATIONAL, Gaithersburg, MD, method **960.46D**, **G**, and **H**.

390   Tryptophan

A. Basal Medium

(a) *Tryptophan-free double-strength basal medium.*—(Stock basal medium solutions, *see* **300F**.)

mL of basal medium to prepare (Add in order listed) (mL)	250	500	1000
Cystine solution [**300F(d)**]	25	50	100
Adenine–guanine–uracil solution [**300F(b)**]	5	10	20
Vitamin solution IV [**300F(o)**]	5	10	20
Vitamin solution VII [**300F(t)**]	5	10	20
Salt solution A [**300F(i)**]	5	10	20
Salt solution B [**300F(j)**]	5	10	20

Add ca 100 mL H_2O and the following solids (g):			
Vitamin-free casein, hydrolyzed	2.5	5	10
Dextrose, anhydrous	10	20	40
Sodium acetate, anhydrous	5	10	20

Swirl to dissolve solids, adjust to pH 6.8 with HCL or NaOH, and dilute to volume with H_2O.

(*Note*: Bacto tryptophan assay medium is no longer available from Difco.)

B. Standard Solutions

(a) *Stock solution (100.0 µg/mL).*—Weigh 50 mg previously dried L-tryptophan into a 500 mL volumetric flask and dissolve with 25% ethanol solution. Dilute to volume with additional 25% ethanol solution.

(b) *Working solutions.*—(Prepare on day of assay.) Dilute 5 mL intermediate solution to 500 mL with H_2O (1.0 µg/mL). This is the primary standard that has an equivalent assay rate to the samples.

Also prepare two secondary standards, one higher (usually 0.8 µg/mL) and one lower (usually 1.4 µg/mL) than the primary standard. These standards are prepared for

55

cases in which the concentration of tryptophan in the sample is lower or higher than that of the primary standard used.

C. Inoculum

(a) *Test organism.*—*Lactobacillus plantarum* (ATCC 8014) maintained through weekly transfers on agar maintenance medium incubated for 16–21 h at 37°C. Store in refrigerator.

(b) *Agar maintenance medium.*—Use **300G(b)**.

(c) *Liquid culture medium.*—Dilute measured volume of basal medium with equal volume of H_2O containing 2.0 μg/mL L-tryptophan standard. Add 10 mL portions of diluted medium to 20×150 screw capped test tubes, autoclave 15 min at 121–123°C, and cool tubes as quickly as possible. Store in refrigerator.

(d) *Working inoculum.*—(Transfer on day before the assay is run.) Transfer cells from *L. plantarum* stock culture to two sterile tubes containing the liquid culture medium, keeping all transfers as sterile as possible. Incubate in a constant temperature bath for 16–21 h at 37°C. Under aseptic conditions, centrifuge the culture and decant the supernatant. Wash cells with 10 mL sterile saline [**300G(c)**]. Repeat procedure two more times. After third washing with sterile saline, resuspend the cells. Cell suspension is the inoculum.

D. Assay Solution

(a) If the concentration of tryptophan in the sample is unknown, then the amount must be estimated prior to analysis. *(1)* The percent moisture (moisture content) and percent protein must be known in order to estimate the miligrams of L-tryptophan.

(2) Using the $Ba(OH)_2$ procedure [**D(b)**], assay for ½ of the total milligrams of tryptophan estimated to be present in the sample. The microorganism responds only to the L-form, therefore, the dilution factor must be multiplied by two to take into account the inactive D-form.

(3) To calculate mg L-tryptophan per unit of sample, remove a sample size (by weight or volume) equivalent to ca 0.3 g protein and use the following formula to obtain mg L-tryptophan per sample aliquot:

$$\text{g protein} \times 1000 \times \frac{0.014}{2} = \text{mg L-tryptophan per sample aliquot}$$

Multiple the miligrams of L-tryptophan per aliquot by the number of aliquots in the sample. Then multiply this number by two to obtain DL-tryptophan and divide by 60 to obtain mg niacin equivalents per sample. If niacin equivalents or DL-tryptophan are not needed, stop at "mg L-tryptophan per sample" and proceed to **D(a)***(5)*.

(4) To calculate the assay, multiply the assay sample value by 30 to take into account the racemization plus the conversion from tryptophan to niacin (60 mg tryptophan is equal to 1 mg niacin). From this point, the serial dilutions and dilution factor are the same as a standard microbiological assay.

(5) If a sample contains a known amount of tryptophan, the serial dilutions and dilution factor calculations are the same as in a standard microbiological assay.

(b) *Ba(OH)$_2$ procedure.*—Remove a sample aliquot containing ca 0.3 g protein and place in a 125 mL polypropylene Erlenmeyer flask. Add 15.4 g Ba(OH)$_2$·8 H$_2$O [or 8.4 g Ba(OH)$_2$ anhydrous] plus an amount of H$_2$O calculated to give a total volume of ca 10 mL H$_2$O per determination when the hydrate is used and ca 17 mL of H$_2$O when the anhydrous form is used. The moisture content of the sample should be corrected for at this stage. Autoclave 30 min at 121–123°C. While still hot, add 50 mL H$_2$O and 15 mL 9 *N* H$_2$SO$_4$ (25% v/v) to each flask. Scrape the bottom and sides of the flask with a glass rod to bring the precipitate into the mixture. Using a stirring bar, adjust the mixture to pH 4.0–4.5 with H$_2$SO$_4$ or NaOH. H$_2$SO$_4$ is added in excess to ensure that all of the Ba^{+2} is combined with SO$_4^{-2}$ to form insoluble BaSO$_4$, eliminating the need to check for complete precipitation in the latter dilution steps. Bring to the first serial dilution steps. Filter, discard the first 50–75 mL cloudy filtrate, and collect the remaining filtrate. Remove an aliquot of the filtrate, add NaOH solution to adjust pH to 6.8, and dilute to final serial dilution volume with H$_2$O. The final volume gives the assay rate concentration of 1.0 µg/mL.

E. Assay

Using standard solution [**390B(b)**], assay solution (**390D**), basal media (**390A**), and inoculum [**390C(d)**], proceed as in **300J** and **K**.

References

(1) Miller, E. L. (1967) *J. Sci. Food Agric.* **18**, 382–386.

(2) *Official Methods of Analysis* (1995) 16th Ed., AOAC INTERNATIONAL, Gaithersburg, MD, method **960.46D**, **G**, and **H**.

(3) Angyal, G. (1985) *Contribution of Tryptophan to Niacin Equivalent of Infant Formula: Microbiological Method*, proceedings of the AOAC Topical Conference on Production, Regulation and Analysis of Infant Formula, May 14–16, 1985, Virginia Beach, VA, pp 152–159.

400 Pyridoxine
(Vitamin B$_6$) AOAC

A. Basal Medium

(a) *Vitamin B$_6$-free double-strength basal medium solutions.*—Store in refrigerator, separate from **300J** stock basal solutions. *(1) Citrate buffer.*—Dissolve 60 g KOH and 82 g citric acid monohydrate in 1 L H$_2$O. *(2) Vitamin solution I.*—Dissolve 10 mg thiamine HCl and 1 g inositol in 200 mL H$_2$O and dilute to 1 L. *(3) Vitamin solution II.*—Dissolve 10 mg biotin in 100 mL 50% ethanol. Dissolve 200 mg D-calcium pantothenate and 200 mg niacin in ca 200 mL H$_2$O, add 8 mL biotin solution, and dilute to 1 L with H$_2$O. *(4) Salt solution I.*—Dissolve 17 g KCl, 10.3 g MgSO$_4$·7H$_2$O, 100 mg FeCl$_3$·6H$_2$O, and 100 mg MnSO$_4$·H$_2$O in ca 800 mL H$_2$O. Add 2 mL concentrated HCl. Dissolve 5 g CaCl$_2$·2H$_2$O in 100 mL H$_2$O, add to first solution, and dilute to 1 L with H$_2$O. *(5) Salt solution II.*—Dissolve 22 g KH$_2$PO$_4$ and 40 g (NH$_4$)$_2$HPO$_4$ in H$_2$O and dilute to 1 L. *(6) Salt solution III.*—Dissolve 4 mg (NH$_4$)$_6$Mo$_7$O$_{24}$·4H$_2$O, 10 mg KI, and 25 mg CuSO$_4$ in H$_2$O and dilute to 100 mL. *(7) Polysorbate 80 solution.*—Weigh 2.5 g polysorbate 80 (Tween 80) and polyoxyethylene sorbitan mono-oleate in small beaker. Dissolve in 45°C H$_2$O and dilute to 500 mL. *(8) Amino acid mixture.*—Weigh 1.1 g D,L-tryptophan, 1.35 g L-histidine·HCl, 5.0 g D,L-methionine, 10.8 g D,L-isoleucine, and 12.8 g D,L-valine. Mix and transfer into bottle with screw cap. Store in desiccator.

mL of basal medium to prepare	250	500	1000
(Add in order listed) (mL)			
Citrate buffer	25	50	100
Vitamin solution I	12.5	25	50
Vitamin solution II	12.5	25	50
Salt solution I	12.5	25	50
Salt solution II	12.5	25	50
Salt solution III	0.25	0.5	1
Polysorbate 80 solution	5	10	20
Add ca 100 mL H$_2$O and the following solids (g):			
Dextrose, anhydrous	2.5	50	100

mL of basal medium to prepare	250	500	1000
(Add in order listed)			
L-asparagine	0.5	1	2
Dissolve the following solid (g) in dilute HCl and add to above solutions:			
Amino acid mixture	0.155	0.31	0.64

Adjust to pH 4.5 with citric acid (1 + 1) and dilute to volume with H_2O.

B. Standard Solutions

(Use dark bottles.)

(a) *Stock solution (10 μg/mL) [Pyridoxine].*—Accurately weigh 12.16 mg USP pyridoxine·HCl dried to constant weight and dissolve in 1 N HCl. Dilute to 1 L with 1 N HCl. Store in refrigerator.

(b) *Intermediate solution (100 ng/mL).*—Prepare day of assay. Dilute 5 mL stock solution to 500 mL with H_2O.

(c) *Working solution.*—Prepare day of assay. Dilute 5 mL intermediate solution to 500 mL with H_2O (1.0 ng/mL). This is the primary standard at the same assay rate as the samples.

Also run two secondary standards, one higher (usually 1.4 ng/mL) and one lower (usually 0.8 ng/mL). These secondary standards are used to bracket the sample concentration when it is too high or too low for the primary standard.

C. Inoculum

(a) *Test organism.*—*Saccharomyces uvarum* (ATCC 9080) maintained through weekly transfers on agar maintenance medium incubated for 16–21 h at 30°C. Store in refrigerator.

(b) *Agar maintenance medium.*—Use Difco wort agar (No. 0111-17) as directed. Heat to boil. Pipet 7 mL hot agar into 20 × 150 mm test tubes, plug with cotton, and cap. Autoclave 15 min at 121–123°C. After autoclaving, tilt hot agar tubes to form slants and cool in position. Store in refrigerator.

(c) *Liquid culture medium.*—Dilute measured volume of basal medium with equal volume of H_2O containing 2.0 ng/mL pyridoxine, pyridoxal, and pyridoxamine. Add 10 mL portions of diluted medium to 20 × 150 mm screw cap test tubes each containing two 4–5 mm glass beads. Autoclave 5 min at 121–123°C and cool tubes rapidly. Store in refrigerator.

(d) *Inoculum rinse.*—Dilute measured volume of basal medium with an equal volume of H_2O. Add 10 mL portions of diluted medium to 20 × 150 mm test tubes and cap. Autoclave 5 min at 121–123°C and cool tubes rapidly. Store in refrigerator.

(e) *Working inoculum.*—Transfer the day before assay is being run. Transfer cell from *S. uvarum* stock culture to two sterile tubes containing the liquid culture medium, keeping all transfers as sterile as possible. Incubate on a rotary shaker (100 rpm) for 16–

21 h at 30°C. Under aseptic conditions, centrifuge culture and decant supernatant. Wash cells with 10 mL sterile inoculum rinse [**C(d)**]. Repeat procedure two more times. Resuspend cells and add 2 mL of cell suspension to 10 mL sterile inoculum rinse. Cell suspension obtained is the inoculum.

D. Assay Solution

(Work in subdued light to prevent loss of B_6.)

(**a**) Accurately weigh a sample containing ≥250 ng vitamin B_6 and transfer into 500 mL Erlenmeyer flask. Limit sample size to ≤2 g.

(**b**) For plant products add 200 mL 0.44 *N* HCl, and for animal products add 200 mL 0.055 *N* HCl. Autoclave plant-base sample for 2 h and animal-base sample for 5 h at 121–123°C and cool.

(**c**) Adjust mixture, with stirring, to pH 6.0–6.5 with KOH; immediately add dilute citric acid until no further precipitation occurs (usually pH 4.5).

(**d**) Dilute mixture to the first volume in the serial dilutions with H_2O and filter.

(**e**) Remove known aliquot of clear filtrate and check for complete precipitation with dilute citric acid. *(1)* If no further precipitation occurs, adjust pH to 4.5 with KOH and dilute with H_2O to the final volume (assay rate is usually 1.0 ng/mL). *(2)* If further precipitation occurs, again adjust the mixture to point of maximum precipitation, dilute with H_2O to a known volume, filter, take known aliquot of clear filtrate, adjust pH to 4.5 with KOH, and dilute with H_2O to the final volume (assay rate is usually 1.0 ng/mL).

E. Assay

Using standard solution [**400B(c)**], assay solution [**400D**], basal media (**400A**), and inoculum [**400C(e)**], proceed as in **300J** and **K** with the following exceptions. Place two 4 mm glass beads into each test tube before adding standard and assay solutions. Sterilize a cloth towel and use to cover test tubes under the stainless steel top before going on shaker for incubation (this prevents any metal fillings from contaminating the assay from the shaking of the test tubes against the metal top). Incubate tubes on constant rotary shaker for 22 h at 100 rpm in a 30°C temperature-regulated room.

References

(1) *Official Methods of Analysis* (1995) 16th Ed., AOAC INTERNATIONAL, Gaithersburg, MD, method **961.15** (modified).

(2) Parrish, W. P., Loy, H. W., and Kline, O. L. (1956) *J. Assoc. Off. Anal. Chem.* **39**, 157–166.

410 Cobalamin

(Vitamin B$_{12}$) AOAC

A. Basal Medium

(a) *Vitamin B$_{12}$-free double-strength basal medium.*—(Solutions, *see* **300F**.) Adjust pH to 6.0 with either HCl or NaOH and dilute to volume with H$_2$O.

	250	500	1000
mL of basal medium to prepare	250	500	1000
(Add in order listed) (mL)			
Adenine–guanine–uracil solution [**300F(b)**]	5	10	20
Asparagine solution [**300F(c)**]	5	10	20
Salt solution A [**300F(i)**]	5	10	20
Salt solution B [**300F(j)**]	5	10	20
Vitamin solution I [**300F(l)**]	10	20	40
Vitamin solution II [**300F(m)**]	10	20	40
Xanthine solution [**300F(r)**]	5	10	20
Polysorbate 80 solution [**300F(h)**]	5	10	20

Add ca 100 mL H$_2$O and the following solids (g):

Vitamin-free casein	2.5	5	10
Dextrose, anhydrous	10	20	40
Sodium citrate·3H$_2$O	8.3	16.6	33.2
Ascorbic acid	1	2	4

Dissolve the following solids (g) in dilute HCl and add to above solution:

L-cystine	0.1	0.2	0.4
D,L-tryptophan	0.1	0.2	0.4

(b) The basal medium can also be prepared by using commercially available Difco B$_{12}$ assay medium USP (No. 0457-15) at a higher assay rate (usually 0.02 ng/mL).

B. Standard Solutions

(Use dark bottles.)

(a) *Stock solution (100 ng/mL).*—Dry USP cyanocobalamin to constant weight. Accurately weigh cyanocobalamin equivalent to 50 µg (directions are provided on USP labels). Dissolve in 25% ethanol. Dilute to 50 mL with additional 25% ethanol and store in refrigerator.

(b) *Intermediate solution (1 ng/mL).*—Dilute 10 mL stock solution to 1 L with 25% ethanol. Store in refrigerator.

(c) *Working solution.*—Prepare day of assay. Dilute 7 mL intermediate solution to 500 mL with H_2O (0.014 ng/mL). This is the primary standard and has the same assay concentration as the sample. In addition, run two secondary standards, one higher (usually 0.02 ng/mL) and one lower (usually 0.01 ng/mL) than the primary standard. This brackets the sample in case the concentration is either higher or lower than the primary standard.

C. Inoculum

(a) *Test organism.*—*Lactobacillus leichmannii* (ATCC 7830) maintained through weekly transfers on agar maintenance medium incubated for 16–21 h at 37°C. Store in refrigerator.

(b) *Agar maintenance medium.*—Use **300G(b)**.

(c) *Liquid culture medium.*—Dilute a measured volume of basal medium with an equal volume of H_2O containing 1 ng vitamin B_{12} per mililiter. Add 10 mL portions of diluted medium to 20 × 150 mm screw cap test tubes, autoclave 15 min at 121–123°C, and cool tubes as quickly as possible. Store in refrigerator. Liquid culture medium can also be prepared by using commercially available Difco B_{12} inoculum broth (USP No. 0542-15) as directed.

(d) *Working inoculum.*—(Transfer the day before assay is being run.) Transfer cells from *L. leichmannii* stock culture to two sterile tubes containing the liquid culture medium, keeping all transfers as sterile as possible. Incubate for 16–21 h in a constant temperature bath at 37°C. Under aseptic conditions, centrifuge culture and decant supernatant. Wash cells with 10 mL sterile saline [**300G(c)**]. Repeat the procedure two more times. After the third washing with sterile saline, resuspend the cells. Dilute aliquot of cell suspension in sterile saline to give percent transmittance equivalent to that of dried cell weight 0.5–0.75 mg per tube (ca 60%T) when read against saline set at 100%T. Cell suspension obtained is inoculum.

D. Assay Solution

(a) Prepare extracting solution just prior to use. Each 100 mL portion contains 1.3 g anhydrous Na_2HPO_4, 1.2 g citric acid, and 1.0 g anhydrous Na metabisulfite ($Na_2S_2O_5$).

(b) Remove quantity of sample containing 50–100 ng vitamin B_{12}. Transfer into a 250 mL beaker containing an amount of extracting solution ≤0.03 mg $Na_2S_2O_5$ per mililiter at a final assay concentration of 0.014 ng/mL vitamin B_{12}.

Example:

$$x \text{ (mL)}$$

$$\text{Sample} \xrightarrow{} A$$
$$B \rightarrow C$$
$$D \rightarrow E \text{ (if necessary)}$$

Where:

$$x = \frac{.03(A)(C)(E)}{10(B)(D)}$$

$$x = \text{mL bisulfite (10 mg/mL)}$$

(c) Disperse sample evenly in the liquid. Wash down sides of beaker with H_2O and cover with watch glass (usually 100 mL total volume).

(d) Autoclave mixture for 10 min at 121–123°C and cool. If lumping occurs, agitate mixture until particles are evenly dispersed.

(e) For samples containing no protein, adjust samples to pH 6.0 with NaOH, dilute to first volume, and filter. For samples containing protein, adjust mixture to pH 4.5 with NaOH while stirring, dilute to first volume, and filter.

(f) Remove known aliquot of clear filtrate and check for complete precipitation of protein by adding dilute HCl. (1) If no further precipitation occurs, adjust to pH 6.0 with NaOH and dilute to final measured volume with H_2O (assay rate is usually 0.014 ng/mL). (2) If further precipitation does occur, again adjust the mixture to maximum precipitation, dilute with H_2O to a known volume, filter, take know aliquot of clear filtrate, adjust to pH 6.0 with NaOH, and dilute to final measured volume with H_2O (assay rate is usually 0.014 ng/mL).

E. Assay

Using standard solution [**410B(c)**], assay solution (**410D**), basal media (**410A**), and inoculum [**410C(d)**], proceed as in **300J** and **K**.

(*Note:* If samples are low using sodium metabisulfite extraction solution, use the cyanide extraction method as follows: prepare 0.1 g KCN or NaCN and 1.2 g NaH_2PO_4 per 100 mL H_2O to use as the extraction solution. Follow the same procedure as above.)

References

(1) *Official Methods of Analysis* (1995) 16th Ed., AOAC INTERNATIONAL, Gaithersburg, MD, method **952.20H**.

(2) Hoff-Jorgensen, E. (1954) *Methods of Biochemical Analysis*, Vol. 1, Interscience Publishers, New York, NY, pp. 81–83.

Appendix A

Standard Solutions and Inoculum Broth

Biotin

Stock solution (100 µg/mL)

Weigh 50 mg Fluka AG, Sigma, or USP biotin dried 3 h at 105°C and cooled in desiccator. Dissolve and dilute to 500 mL in 25% ethanol. Store in refrigerator.

Intermediate solution (1 µg/mL)

Dilute 5 mL stock solution to 500 mL with 25% ethanol. Store in refrigerator.

Inoculum broth

Add 250 mL H_2O containing 2 ng/mL biotin to 250 mL basal medium. Pipet 10 mL into each test tube and autoclave 15 min at 121°C. Store in refrigerator.

Panthenol (Pantothenol or Pantothenyl Alcohol)

Stock solution (50 µg/mL) (pantoic acid)

Dissolve 69.2 mg pure Panthenol in distilled H_2O, adjust to pH 6.0, and dilute to 1 L with H_2O. One mililiter contains the equivalent of 50 µg pantoic acid. Store under toluene in refrigerator

Inoculum broth

Add 10 mL Difco Panthenol supplement and 2 µg pantoic acid to 190 mL Difco Panthenol inoculum broth. (Pipet 10 mL stock solution and 10 mL 0.1 *N* NaOH into a 125 mL Erlenmeyer flask and autoclave 30 min. Cool, adjust to pH 6.0 with 0.2 *N* HCl, and diluted to 250 mL with H_2O). Pipet 10 mL into each test tube and autoclave 10 min at 100°C. Store in refrigerator.

Maintenance medium

Add 5 g mannitol, 0.5 g yeast extract, 0.5 g peptone, and 1.5 g agar to 100 mL H_2O. Heat until agar melts, pipet 7 mL into each test tube, and autoclave 15 min at 121°C. Cool at slant and store in refrigerator.

Pantothenic Acid

Stock solution (40 µg/mL)
Weigh 43.47 mg USP calcium pantothenate that has been dried for 3 h at 105°C and cooled in desiccator. Dissolve in ca 500 mL H_2O. Add 10 mL 0.2 N acetic acid and 100 mL 0.2 N sodium acetate. Dilute to 1 L with H_2O. Store under toluene in refrigerator.

Intermediate solution (1 µg/mL)
Add ca 500 mL H_2O to 25 mL stock solution. Add 10 mL 0.2 N acetic acid and 100 mL 0.2 N sodium acetate. Dilute to 1 L with H_2O. Store under toluene in refrigerator.

Inoculum broth
Dilute a measured volume of basal medium with an equal volume of H_2O containing 0.04 µg pantothenate per mililiter. Pipet 10 mL into each test tube and autoclave 15 min at 121°C. Store in refrigerator.

Niacin (Nicotinic Acid)

Stock solution (100 µg/mL)
Weigh 50 mg USP niacin that has been dried 1 h at 105°C and cooled in desiccator. Dissolve in 25% ethanol and dilute to 500 mL with additional 25% ethanol.

Intermediate solution (1 µg/mL)
Dilute 5 mL of stock solution to 500 mL with 25% ethanol.

Inoculum broth
Dilute measured volume of basal medium with equal volume of H_2O containing 0.2 µg niacin per mililiter. Pipet 10 mL into each test tube and autoclave 15 min at 121°C. Store in refrigerator.

Tryptophan

Stock solution (1 mg/mL)
Weigh 500 mg USP L-tryptophan that has been dried 3 h at 105°C and cooled in desiccator. Dissolve and then dilute to 500 mL with 25% ethanol.

Intermediate solution (100 µg/mL)
Dilute 50 mL stock solution to 500 mL with 25% ethanol.

Inoculum broth
Mix 250 mL basal medium with an equal amount of H_2O containing 2 µg L-tryptophan per mililiter. Pipet 10 mL into each test tube and autoclave 15 min at 121°C. Store in refrigerator.

Folic Acid (Pteroylglutamic Acid)

Stock solution (100 μg/mL)

Dissolve 50 mg USP folic acid that has been dried to constant weight in ca 30 mL 0.01 *N* NaOH (0.4 g NaOH per liter). Add 300 mL H_2O, adjust to pH 7–8 with HCl, and dilute to 500 mL with H_2O. Store solution in a dark bottle under toluene in refrigerator.

Intermediate solution 1 (1 μg/mL)

Add ca 500 mL H_2O to 5 mL stock solution, adjust pH to 7–8 with HCl and dilute to 1 L with H_2O. Store solution in a dark bottle under toluene in refrigerator.

Intermediate solution 2 (10 ng/mL)

Add ca 500 mL H_2O to 10 mL intermediate solution 1, adjust pH to 7–8 with HCl and dilute to 1 L. Store solution in dark bottle under toluene in refrigerator.

Inoculum broth

S. faecalis, use Difco B_{12} inoculum broth. Pipet 10 mL into each test tube and autoclave 15 min at 121°C. Store in refrigerator.

L. casei, mix 250 mL Difco folic acid basal medium with an equal amount of H_2O containing 0.2 μg folic acid and 5 mg solubilized liver extract. (Prepare liver extract by suspending 5 g liver in 100 mL H_2O. Hold mixture for 1 h at 50°C and then for 5 min at 80°C and filter. The filtrate contains ca 8 mg solid per liter. Store under toluene in refrigerator.) Pipet 10 mL into each test tube, autoclave for 15 min at 121°C, and store in refrigerator. Difco bacto micro inoculum broth (No. 0320-02) has also been found satisfactory.

Vitamin B_6 (Pyridoxine)

Stock solution (10 μg/mL)

Weigh 12.16 mg USP pyridoxine·HCl dried to constant weight and dissolve in 1 *N* HCl (83.3 mL HCl). Dilute to 1 L with 1 *N* HCl. Store in dark bottle in refrigerator.

Intermediate solution (100 ng/mL)

Dilute 5 mL stock solution to 500 mL with 1 *N* HCl.

Liquid culture stock solution (10 μg/mL)

Prepare separate solutions of 12.16 mg pyridoxine·HCl, 12.18 mg pyridoxal·HCl, and 14.34 mg pyridoxamine·2HCl each diluted to 1 L in 1 *N* HCl. Store in refrigerator.

Liquid culture intermediate solution (1 μg/mL)

Dilute 10 mL of each liquid culture stock solution to 100 mL with H_2O. Prepare fresh as needed.

Mixed liquid culture solution

Pipet 1 mL of each liquid culture intermediate solution into a 500 mL volumetric flask and dilute to volume with H_2O. Prepare fresh as needed.

Inoculum broth

Pipet 5 mL mixed liquid culture solution into test tube containing 2 glass beads. Add 5 mL basal medium and autoclave 5 min at 121°C. Store in refrigerator.

Wort agar slants

Dissolve 25.3 g Difco wort agar and 3 g Difco agar in 500 mL H_2O. Bring to boil and pipet 7 mL into each test tube. Plug with cotton and cap. Autoclave 15 min at 121°C and cool slanted.

Vitamin B_{12} (Cyanocobalamin)

Stock solution (100 ng/mL)

Dissolve a dried, constant weight aliquot of USP cyanocobalamin equivalent to 50 μg vitamin B_{12} in 25% ethanol. Dilute to 500 mL with additional 25% ethanol and store in dark bottle in refrigerator.

Intermediate solution (1 ng/mL)

Dilute 5 mL stock solution to 500 mL with 25% ethanol.

Liquid culture stock solution (1 μg/mL)

Dissolve a dried, constant weight aliquot of USP cyanocobalamin equivalent to 250 μg vitamin B_{12} in 25% ethanol. Dilute to 250 mL with additional 25% ethanol.

Inoculum broth

Dissolve 15 g peptonized milk, 5 g yeast extract, 10 g dextrose, and 2 g KH_2PO_4 in ca 600 mL H_2O. Add 100 mL filtered tomato juice and adjust to pH 6.5–6.8 with NaOH. Add, with mixing, 10 mL polysorbate 80 solution (25 g Tween 80 to 250 mL H_2O) and 1 mL liquid culture stock solution, dilute to 1 L with H_2O. Pipet 10 mL into each test tube and autoclave 15 min at 121°C. Store in refrigerator. Difco B_{12} inoculum broth USP can be substituted.

Vitamin B_1 (Thiamine)

Stock solution (100 μg/mL)

Weigh 50 mg USP thiamine·HCl that has been dried to constant weight and dissolve in 20% ethanol (adjusted to pH 3.5–4.3 with HCl). Dilute to 500 mL with acidified ethanol and store in dark bottle in refrigerator.

Intermediate solution (10 μg/mL)

Dilute 50 mL stock solution to 500 mL with 20% ethanol adjusted to pH 3.4–4.3 with HCl. Use dark bottle to store in refrigerator.

Inoculum broth

Dissolve 23.1 g Difco APT broth (No. 0655-01-7) and 2.5 mL (100 µg/mL) thiamine standard in 500 mL H_2O. Pipet 10 mL into each test tube and autoclave 15 min at 121°C. Store in refrigerator.

Maintenance medium

Dissolve 30.6 g Difco APT agar (No. 0654-01-8) in H_2O and bring to boil. Add 500 µg thiamine to solution and dilute to 500 mL with H_2O. Pipet 10 mL into each test tube and autoclave 15 min at 121°C. Store in refrigerator.

Vitamin B₂ (Riboflavin)

Stock solution (100 µg/mL)

Weigh 50 mg USP riboflavin that has been dried 2 h at 105°C and cooled in desiccator. Suspend in ca 300 mL 0.02 *N* acetic acid (1.2 mL acetic acid in 1 L H_2O). Warm on a steam bath, with stirring, until solid dissolves. Cool and dilute to 500 mL with 0.02 *N* acetic acid. Store in dark bottle under toluene in refrigerator.

Intermediate solution (1 µg/mL)

Dilute 5 mL stock solution to 500 mL with 0.02 *N* acetic acid. Store in dark bottle under toluene in refrigerator.

Inoculum broth

Dilute measured volume of basal medium with equal volume of H_2O containing 0.1 µg riboflavin per mililiter. Pipet 10 mL into each test tube and autoclave 15 min at 121°C. Store in refrigerator.

Choline

Stock solution (500 µg/mL)

Accurately weigh 576 mg choline chloride that has been dried to constant weight. Dissolve in H_2O and dilute to 1 L. Store in refrigerator.

Inoculum

Transfer cells from *Neurospora crassa* (ATCC 9277) stock culture to a sterile saline test tube. Suspension obtained is the inoculum.

Appendix B
Time Factors

DAY 1

(a) Sample calculations (serial dilutions and dilution factors).

(b) Transfer samples from plugs or slants into inoculum broth for specific vitamin to be analyzed.

(c) If totals are being run, set up the enzymatic procedure.

DAY 2

(a) If using the enzymatic procedure, autoclave the samples and standards 3 min at 100°C to deactivate the enzymes. Then proceed as in standard microbiological assay without the autoclaving step(s).

(b) *Standard microbiological assay.*—Weigh, add extracting fluid, autoclave, and cool sample. Make basal medium (from scratch or use Difco) Place test tubes in rack. Adjust the sample pH, filter, and serial dilute to assay rate (same as primary standard). Pipet samples and standards into test tubes (final volume 5 mL). Add 5 mL basal medium to all test tubes for a total of 10 mL. Autoclave 5 min at 121°C and cool. Spin down inoculum at 1500–2000 rpm, decant and discard liquid, and rinse solid with 10 mL of sterile saline (0.9% NaCl). This is repeated two more times (with the exception of B_6, which has its own inoculum rinse). Inoculate all tubes (except uninoculated blank) with one drop (see individual assays for exceptions). Place in H_2O bath for 16–24 h at prescribed temperatures.

DAY 3

(a) Read check tubes at 2 h increments until assay is complete

(b) Remove assay from H_2O bath and allow to come to room temperature.

(c) Turn on reading instrument and warm-up for a minimum of 30 min.

(d) Read tubes.

(e) Calculate results.

Appendix C
Troubleshooting

Complete absence of growth in all tubes

(**a**) Some ingredient missing from the basal medium.

(**b**) Deterioration of some constituent of the medium.

(**c**) Medium at incorrect pH.

(**d**) Some inhibitory substance (e.g., detergent in medium or glassware).

(**e**) Mutation of test organism.

(**f**) Inoculation with the wrong microorganism.

(**g**) Overnight change in the temperature of the incubator.

(**h**) Tubes inoculated before they have cooled sufficiently.

(**i**) Failure to inoculate tubes.

Absence of growth in samples only

(**a**) Sample contains antibiotics or other inhibitor.

(**b**) Wrong dilutions chosen.

(**c**) Failure to inoculate sample tubes.

Absence of growth in the standard only

(**a**) Deterioration of the standard (such as from exposure from light).

(**b**) Standard kept too long.

(**c**) Inhibitory material introduced into the stock solution.

(**d**) Some constituent missing from medium but by chance present in the sample.

(**e**) Failure to inoculate standard tubes.

Excessive growth in all tubes, including the inoculated blanks

(**a**) The assayed vitamin or a substitute is present in the medium.

(**b**) The assayed vitamin or a substitute is present in the glassware or diluent.

(**c**) Carry-over of assayed vitamin from the transfer medium (culture medium).

(**d**) Mutation of test microorganism. If microorganisms are maintained on a medium with insufficient nutrient, they may mutate and develop the ability to grow without it.

Irregular growth

(**a**) Insufficiently trained worker setting up the assay.
(**b**) Dirty glassware.
(**c**) Loss of activity of a constituent of the medium.
(**d**) Contamination with other bacteria.
(**e**) Assay vitamin in or on tube closures.
(**f**) Tubes allowed to stand in bright sunlight (folic acid).
(**g**) Carry-over from the culture medium.
(**h**) Organisms undergoing mutation.
(**i**) Different temperatures in different areas of incubator.
(**j**) Vibrations affecting incubator erratically.
(**k**) Faulty photoelectric spectrophotometer.
(**l**) Saliva contaminating tubes during pipetting.
(**m**) Valve sticking in automatic mechanical pipet.

Excessively high or low results

(**a**) Miscalculations.
(**b**) Deterioration of standard.
(**c**) Sample potency different than expected.
(**d**) Omission or addition of a dilution step.

Appendix D
Assay Reference Guide

Assay	Organism	Basal medium	Assay concentration, ng/mL	Standard concentration, ng/mL	pH	Extraction solution	Autoclave time, min	Special Notes
Thiamine (vitamin B$_1$)	L. viridescens	Difco	4	3 4 5	6.0	0.1 N HCl	30	Inoculum 1:10; incubate 18 h
Riboflavin (vitamin B$_2$) AOAC	L. casei	Difco	30	20 30 40	6.8	0.1 N HCl	30	Keep pH below 7
Pyridoxine (vitamin B$_6$) AOAC	S. uvarum	Modified AOAC	1	0.8 1 1.4	4.5	0.44 or 0.055 N HCl	120 or 350	Inoclum 2:10 incubate 22 h
Cyanoco-balamin (vitamin B$_{12}$) AOAC	L. leichmannii	AOAC	0.014	0.01 0.014 0.02	6.0	Bisulfate or cyanide	10	Inoculum dropwise to 60%T
Biotin	L. plantarum	Difco	0.1	0.08 0.1 0.14	6.8	1 N H$_2$SO$_4$	30	Inoculum as is
Niacin AOAC	L. plantarum	AOAC	10	8 10 14	6.8	1 N H$_2$SO$_4$	30	Inoculum as is
Tryptophan	L. plantarum	Modified AOAC	1000	800 1000 1400	6.8	Ba(OH)$_2$ plus 17 mL H$_2$O	30	Inoculum as is

Continued

Assay Reference Guide continued

Assay	Organism	Basal medium	Assay concentration, ng/mL	Standard concentration, ng/mL	pH	Extraction solution	Autoclave time, min	Special Notes
Choline	N. crassa	Difco	5000	3000 5000 7000	5.5	3% H_2SO_4	120	Inoculum loop stock culture in saline
Folic acid AOAC	S. faecalis	AOAC	1	0.8 1 1.4	6.8	2 mL NH_4OH (2 + 3) per 100 mL	15	Inoculum as is
Folic acid	L. casei	Difco	0.2	0.12 0.2 0.3	6.8	Na ascorbate buffer	15	Inoculum 5 drops to 10 mL
Total folates AOAC modified	L. casei	Difco	0.2	0.12 0.2 0.3	6.8	Na ascorbate buffer	15	Chicken pancreas enzyme
Pantothenic acid AOAC	L. plantarum	AOAC	10	8 10 14	6.8	Acetic acid and sodium acetate, pH 5.6 buffer	7	Inoculum as is; keep pH below 7
Total pantothenates	L. plantarum	AOAC	10	8 10 14	6.8	Tris buffer	5	Pigeon liver extract intestinal phosphatease enzymes
Panthenol	A. suboxy-dans	Difco	400	300 400 500	6.0	H_2O then 0.1 N NaOH	30	Inoculum as is, keep below pH 5

Appendix E

Stability and Lability of Vitamins

Nutrient	Neutral	Less than pH 7	Greater than pH 7	O_2	Light	Heat
Biotin	S	S	S	S	S	S
Choline	S	S	S	L	S	S
Cobalamin	S	S	S	L	L	S
Folic acid	L	L	S	L	L	S
Tryptophan	S	S	S	S	S	S
Niacin	S	S	S	S	S	S
Pantothenic acid	S	L	L	S	S	L
Pyridoxine group	S	S	S	S	L	S
Riboflavin	S	S	L	S	L	S
Thiamine	L	S	L	L	S	L

S = stable   L = labile

Agriculture and Food Chemistry (1959) **7**, 94.

Appendix F
Measuring Recovery Using a Spiked Sample

Spiked samplese are used to demonstrate that an assay is valid for a particular vitamin or other analyte by measuring recovery of a known amount of the analyte from a sample matrix. (The calculated recovery of the analyte is the fraction added to the sample prior to analysis that is measured by the method.)

(1) The concentration of the analyte added to the sample (the spike) should not be less than the concentration of the analyte added to the unspiked sample. The sum of the concentration of added analyte plus the analyte present before spiking should be within the same range as the analyte concentration found in actual samples. Adding the spike must not cause the final expected concentration to exceed the linear range of the standard curve. To avoid bias, both spiked and unspiked samples must be treated identically during analysis.

(2) The percent recovery of analyte from a spiked sample is calculated as follows:

$$\%R = [(C_F - C_U/C_A] \times 100$$

where

   R = recovery

   C_F = the concentration of analyte measured in the spiked sample

   C_U = the concentration of the analyte measured in the unspiked sample

   C_A = the concentration of analyte added to the spiked sample. (Note: It is a calculated value and not a measured value.)

(3) Example of calculation of recovery

Analyzed concentration of analyte in sample:	120 µg/g
Spike added to sample at same concentration found on first analysis:	120 µg/g
Analyzed concentration of analyte in spiked sample:	250 µg/g

Calculation

$$\% \text{ Recovery} = [(250 - 120)/120] \times 100 = 108.3\%$$

Reference

(1) *Official Methods of Analysis* (1995) 16th Ed., AOAC INTERNATIONAL, Gaithersburg, MD, p. xviii.

Appendix G

Turbidimetric Microbiological Assay Program

A. Enter Data

The program requests general assay data, standards data, and sample data. An opportunity to change the standards and sample data is provided in case some of the data is entered incorrectly.

General data

General Data	Allowable range	Variable
Number of samples	1–60	N
Number of standards	1–5	M
Units common to all assays	—	$A\$$
Analyst	—	$B\$$
Assay date	—	$C\$$
Assay organism	—	$D\$$
Incubation time	—	$E\$$

Standards data loop

Do M times ($I = 1$ to M, $J = 1$ to 15) (enter standards in order of increasing concentration and enter percent transmittance values for lowest to highest level of added vitamin).

Concentration factor for standard	$M(I)$
Standard identification	$ID\$(I)$
Percent transmittance	$PT(I,J)$

Samples data loop

Do N times ($I = 1$ to N, $J = 1$ to 8) (enter percent transmittance values for lowest to highest level of added vitamin).

Sample number	$SN\$(I)$
Dilution factor	$DF(I)$
Percent transmittance	$ST(I,J)$

B. Calculate Standard Curves

The program creates standard curves from standard data (percent light transmittance versus milliliters of added vitamin).

Each standard curve is calculated using the 15 values for percent transmittance of a given standard,

$PT\,(I,1)$	$PT\,(I,2)$	$PT\,(I,3)$
$PT\,(I,4)$	$PT\,(I,5)$	$PT\,(I,6)$
$PT\,(I,7)$	$PT\,(I,8)$	$PT\,(I,9)$
$PT\,(I,10)$	$PT\,(I,11)$	$PT\,(I,12)$
$PT\,(I,13)$	$PT\,(I,14)$	$PT\,(I,15)$

where I varies from 1 to M (where M is the value entered for the number of standards used). Each set of 15 values forms a 3×5 array (five levels of vitamin added performed in triplicate).

The first three replicate values are averaged, followed by the second three replicate values, etc. If any value is 0, the average of the non-zero values is calculated.

Each set of 15 values results in a 6-point standard curve plotting percent light transmittance versus milliliters added vitamin. Five of the points are calculated from the data [$SC(I,1)$ to $SC(I,5)$], and the sixth point is 100% transmittance for 0 mL of added vitamin.

The cutoff points equivalent to 0.5 and 4.5 mL of added vitamin [$MN(I)$ and $MX(I)$] are also calculated. The cutoff for 0.5 is the average of 100 % transmittance and the average value calculated for the lowest level of added vitamin [$SC(I,1)$]. The cutoff for 4.5 is the average of the two average values calculated for the two highest levels of added vitamin [$SC(I,4)$ and $SC(I,5)$].

C. Calculate Sample Values

The program takes sample data, selects appropriate standard curve, and eliminates values that fall outside the 0.5 to 4.5 mL of added vitamin cutoff limits. It uses the selected standard curve to convert percent transmittance data to milliliters of added vitamin data and normalizes values to a per milliliter basis. It then calculates the average of normalized values, determines 10% deviation limits around the average, calculates average of values that remain within the 10% deviation limits, and then calculates the amount of vitamin in the sample.

Select standard curve

The eight values for percent transmittance of a given sample form a 2×4 array (four levels of vitamin added performed in duplicate).

$$ST(I,1) \qquad ST(I,2)$$
$$ST(I,3) \qquad ST(I,4)$$
$$ST(I,5) \qquad ST(I,6)$$
$$ST(I,7) \qquad ST(I,8)$$

where I varies from 1 to N (where N is the value entered for the number of samples).

The percent transmittance values for the highest level [$ST(I,7)$ and $ST(I,8)$] are used to select the appropriate standard curve. The values are compared to the cutoff points equivalent to 4.5 mL of added vitamin for the various standard curves [$MX(1)$ to $MX(M)$]. If both values are lower than the cutoff point for the lowest standard curve [$MX(1)$] then the lowest standard curve is selected. If either value is greater than $MX(1)$, then the cutoff point for the next higher standard curve [$MX(2)$] is checked. The process is repeated until both $ST(I,7)$ and $ST(I,8)$ fall within 4.5 mL of some standard curve and that standard curve is selected, or until it is determined that the sample values are high for all standard curves, in which case the highest standard curve is selected. The number of the standard curve selected is assigned to the variable SC.

Adjust data set so that all values fall between 0.5 and 4.5 on selected standard curve

The assay is valid between 0.5 and 4.5 mL of added vitamin on a given standard curve. The cutoff points for the standard curve [$MN(I)$ and $MX(I)$] are used to determine which of the sample values can be evaluated using the selected standard curve. A second data set is created [$ST(I,9)$ to $ST(I,16)$], which is identical to the first except that values outside of the cutoff points for the standard curve are set to zero.

Convert sample data for percent transmittance to "equivalent volume"

The 6-point standard curve of percent light transmittance versus milliliters of added vitamin is used to convert the percent transmittance of light for the sample data to milliliters of added vitamin, the equivalent volume. The conversion is achieved by linear interpolation between two points on the standard curve. A third data set is created [$ST(I,17)$ to $ST(I,24)$], which contains the equivalent volume values for all non-zero data points. Data points that have been set to zero are transferred as zeros.

Normalize "equivalent volume" values

The equivalent volume values are normalized by dividing by the nominal volume of vitamin added (i.e., the first two are divided by one, the second two are divided by two, etc.). A fourth data set is created [$ST(I,25)$ to $ST(I,32)$], which contains the normalized values.

Calculate average of tubes in fourth data set

The average of the nonzero values in the fourth sample data set is determined [*AV(I,1)*]. This is the average normalized equivalent volume for those data values that lie within the 0.5 to 4.5 cutoff limits. Ten percent of this average is added to and subtracted from the calculated average. The two resulting values [*R1(I)* and *R2(I)*] are the two extremes of the range of values that represent a 10% deviation from the average.

Determine which tubes fit within the 10% limits

The values for the range [*R1(I)* and *R2(I)*] are used to determine which of the normalized values lie within the 10% deviation limits. A fifth data set is created [*ST(I,33)* to *ST(I,40)*], which is identical to the fourth data set except that values outside of the 10% deviation limits are set to zero.

Calculate average of tubes in fifth data set

The average of the non-zero values in the fifth data set is determined [*AV(I,2)*]. This is the average, normalized equivalent volume for those data values that lie within the 0.5 to 4.5 mL cutoff limits and the 10% deviation limits.

Calculate answer

The answer [*AA*] is determined as the product of the average, normalized equivalent volume of the tubes within the 0.5 to 4.5 mL cutoff limits and the 10% deviation limits [*AV(I,2)*], the dilution factor for the sample [*DF(I)*], and the concentration factor for the standard used [*M(I)*].

$$AA = AV(I,2) \times DF(I) \times M(I)$$

D. Print Results to Printer

For each sample, general data, standard data, sample data, and calculated results are printed out on the printer. The general data include the analyst's name, *B$*, the assay date, *C$*, the assay organism, *D$*, the incubation time, *E$*, and the units for the answer that are common to all assays, *A$*. The standard data printed include the identification for the standard used, *ID$(I)*, and the percent transmittance data for that standard, *PT(I,J)*. The sample data printed include the sample number, *SN$(I)*, the dilution factor, *DF(I)*, and the percent transmittance data for the sample, *ST(I,J)*, where *J* = 1 to 8. The calculated values printed include the number of tubes used to calculate the answer, *D*, the average normalized "equivalent volume" for those data values that lie within the 0.5 to 4.5 cutoff limits, *AV(I,1)*, the two extremes of the range of values that represent a 10% deviation from the average, *R1(I)* and *R2(I)*, the average, normalized "equivalent volume" for those data values that lie within the 0.5 to 4.5 mL cutoff limits and also the 10% deviation limits, *AV(I,2)*, the eight values for "equivalent volume" that fall within the 0.5 to 4.5 mL cutoff limits and also the 10% deviation limits rounded to three places, *ST(I,J)* where *J* = 41 to 48, and the answer, *AA*.

Symbols

D	Counter (number of non-zero values)
I, J, K	Incremental counters for loops
M	Number of standards (Value between 1 and 5 inclusive)
N	Number of samples (Value between 1 and 60 inclusive)
AA	Answer
II	Incremental counter for loops within subroutines
MQ	"Equivalent volume" value (value from subroutine)
PT	Percent transmittance value (value for subroutine)
S7, S8	2% transmittance values for sample for the highest level of added vitamin (values for subroutine)
SC	Number of the standard curve selected for a given sample (value from subroutine)
XX, YY, ZZ	Summation variables for calculations
M(I)	Concentration factor for standard ($I = 1$ to M)
DF(I)	Dilution factor for sample ($I = 1$ to N)
MN(I), MX(I)	Percent transmittance values for 0.5 and 4.5 mL of added vitamin for standard ($I = 1$ to M)
R1(I), R2(I)	Lower and upper range values for 10% deviation limits rounded to three decimal places ($I = 1$ to N)
AV(I,J)	($J = 1$) Average of normalized "equivalent volume" values within 0.5 to 4.5 mL limits
	($J = 2$) Average of normalized "equivalent volume" values within 0.5 to 4.5 mL limits and also within 10% deviation limits ($I = 1$ to N) ($J = 1$ to 2)
PT(I,J)	Percent transmittance values for standard ($I = 1$ to M) ($J = 1$ to 15)
SC(I,J)	Calculated values for standard curves ($I = 1$ to M) ($J = 1$ to 5)
ST(I,J)	Percent transmittance values, calculated values, and adjusted values for sample ($I = 1$ to N) ($J = 1$ to 48)
	($J = 1$ to 8) Percent transmittance values for sample
	($J = 9$ to 16) Percent transmittance values between 0.5 and 4.5 mL of added vitamin limits on selected standard curve

(J = 17 to 24) "Equivalent volume" values between 0.5 and 4.5 mL of added vitamin limits

(J = 25 to 32) Normalized "equivalent volume" values between 0.5 and 4.5 mL added vitamin limits

(J = 33 to 40) Normalized "equivalent volume" values between 0.5 and 4.5 mL added vitamin and also within 10% deviation limits

(J = 41 to 48) Normalized "equivalent volume" values between 0.5 and 4.5 mL added vitamin and also within 10% deviation limits rounded to three decimal places

A$	Units common to all assays
B$	Analyst's name
C$	Assay date
D$	Assay organism
E$	Incubation time
Q$	Answer to questions
S$	One space
S5$	Five spaces
ID$(I)	Standard identification (I = 1 to M)
SN$(I)	Sample number identification (I = 1 to N)

Reference

(1) Anderson, E. M. (1989) *J. Assoc. Off. Anal. Chem.* **72**, 950–952.

Appendix H

Limits of Quantification

Component	Approximate limits of quantification	
	g/100 g	g/1 g
Biotin	1.5	0.015
Vitamin B_{12}	0.15	0.0015
Vitamin B_6	10	0.10
Niacin	150	1.5
Thiamine	50	0.05
Riboflavin	100	1.0
Pantothenic acid	200	2.0
Folic acid	5	0.05
Tryptophan	2000	20.0

Limit of quantification is the level of analyte in the test sample that produces a signal sufficient to allow the determination of the analyte at least 95% of the time.